# F

# GUIDE TO THE
# HUMAN
# BODY

## RICHARD WALKER

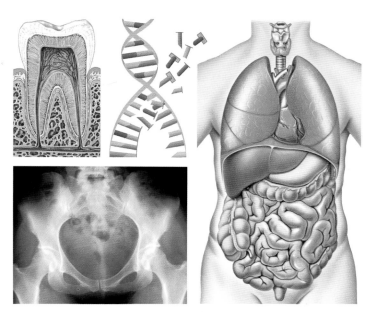

FIREFLY BOOKS

# A FIREFLY BOOK

Published by Firefly Books Ltd. 2004

Copyright © 2004 Philip's

Second printing, 2006

Publisher Cataloging-in-Publication Data (U.S.)

Walker, Richard, 1951–
    Guide to the human body / Richard Walker
1st ed [192]p. : col.ill., col. photos. ; cm.
Includes index.
Summary: Illustrated guide to the human body covering anatomy and physiology. Including a medical encyclopedia of medical conditions and elements of human biology.
ISBN-13: 978-1-55297-879-5 (pbk.)
ISBN-10: 1-55297-879-6 (pbk.)
1. Human physiology – Popular works. 2. Human anatomy – Popular works. 3. Body, Human – Popular works. I. Title.
612 21    QP38.W35 2004

National Library of Canada Cataloguing in Publication Data

Walker, Richard, 1951-
Guide to the human body / Richard Walker.
Includes index.
ISBN-13: 978-1-55297-879-5
ISBN-10: 1-55297-879-6
1. Human anatomy--Popular works. 2. Human physiology--Popular works.
3. Medicine, Popular. I. Title.
QP38.W35 2004        612        C2003-904966-3

Published in the United States by
Firefly Books (U.S.) Inc.
P.O. Box 1338, Ellicott Station
Buffalo, New York 14205

Published in Canada by
Firefly Books Ltd.
66 Leek Crescent
Richmond Hill, Ontario L4B1H1

Printed in China

COMMISSIONING EDITOR  Christian Humphries

EDITOR  Joanna Potts

EXECUTIVE ART EDITOR  Mike Brown

DESIGNER  Caroline Ohara

PRODUCTION  Sally Banner

FRONT COVER (CLOCKWISE):
left:            Cross-section of tooth (Philip's)
centre:        Double helix of DNA  (Philip's)
bottom right: Digestive system (Philip's)
bottom left:  X-ray of a female adult pelvis
                (Science Photo Library)

BACK COVER:
left:        Horizontal section through the eye (Philip's)
right:       Skull (Philip's)

## PHOTOGRAPHY CREDITS

All photographs from Science Photo Library
8c L. Willatt, East Anglia Genetics; 19bl James Stevenson; 19br Dave Roberts; 21br Mehau Kulyk; 23tr Hugh Turvey; 23bl Bo Veisland, MI&I; 23br James Stevenson; 24tr James Watney; 25tl James Stevenson; 25bl Bo Veisland, MI&I; 25br Astrid & Hanns-Frieder Michler; 33tr BSIP, Sercomi; 35br Don Fawcett; 38br Scott Camazine; 41tr Francis Leroy, Biocosmos; 43tl SPL; 49tr Susumu Nishinaga; 61 Russell Kightley; 65 Francis Leroy, Biocosmos; 69tr CNRI; 72bl Biology Media; 73t CNRI; 73b CNRI; 75 CNRI; 109 Mike Bluestone; 112 NIBSC; 113 Alfred Pasieka; 115 Alfred Pasieka; 117 A.B Dowsett; 120 Sidney Moulds; 122 Barry Dowsett; 123 CNRI; 126 Alfred Pasieka; 131 Simon Fraser, Royal Victoria Infirmary, Newcastle upon Tyne; 132 SPL; 135 Alfred Pasieka; 137 John Bavosi; 139 BSIP Ducloux / Brisou; 142 John Bavosi; 144 John Bavosi; 145 Dr Andrejs Liepins; 146 NIBSC; 147 Lawrence Lawry; 149 Dr Steve Patterson; 150 John Bavosi; 152 Claude Nuridsany & Marie Perennou; 153 SPL; 154 John Bavosi; 156 NIBSC; 157 Simon Fraser, Royal Victoria Infirmary, Newcastle upon Tyne; 158 John Bavosi; 161 Eye of Science; 162 Dave Roberts; 163 SPL; 164 SPL; 165 Sinclair Stammers; 166 Alfred Pasieka; 167 CDC; 168 Wellcome Dept. of Cognitive Neurology; 169 Sidney Moulds; 170 Mehau Kulyk; 171 BSIP VEM; 172 Biophoto Associates; 173 Dr Linda Stannard, UCT; 174 Eye of Science; 175 CNRI; 176 Eye of Science; 177 SPL; 180 Sidney Moulds; 181 CNRI; 182 Scott Camazine; 186 SPL; 187 Eye of Science.

# Contents

# THE STRUCTURE OF THE BODY

The human body is a living structure of incredible complexity. The purpose of this book is to describe simply yet comprehensively the anatomy (structure), physiology (function), and interdependence of the body's component parts. Throughout the book, for ease of description, specific terms are used to describe different regions of the body, and the orientation and position of body parts. This terminology, in common usage by doctors and scientists, is explained below.

## Body regions

When viewed externally, the whole body is divided into regions, or areas. The head houses the brain and major sense organs. It is supported and protected by the brain, which also forms the framework of the face. The head is held upright by the muscles and bones of the neck,

which connects the head to the trunk. The **trunk** (or torso) forms the central part of the body and has two sections: the **thorax** forms the upper part of the trunk and extends from the neck to the diaphragm. The **diaphragm** is a sheet of muscle that separates the thorax from the abdomen, the lower part of the trunk. The terms cephalic, cervical, thoracic, and abdominal describe items found, respectively, in the head, neck, thorax, or abdomen. The two upper **limbs** (or extremities) are each divided into three regions: the arm, forearm, and hand; the hollow just beneath the junction between the upper extremity and trunk is the **axilla** (or armpit). The two lower limbs (or extremities) are each divided into the thigh, leg, and foot. Most organs, such as the heart and stomach, are enclosed inside one of three closed cavities within the body. Females and males have the

**Body regions**

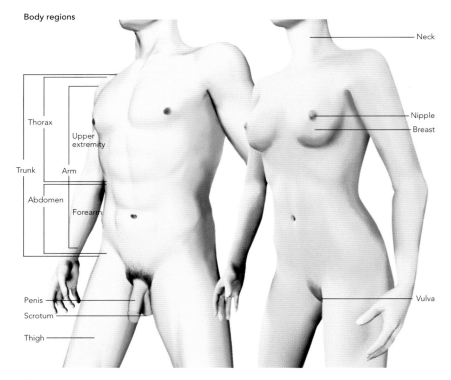

- Thorax
- Upper extremity
- Trunk
- Arm
- Abdomen
- Forearm
- Penis
- Scrotum
- Thigh
- Neck
- Nipple
- Breast
- Vulva

4

same body regions, but their body shapes, and internal and external reproductive organs, differ.

## Orientation and direction

The terminology that describes orientation and direction assumes that the body is upright, with arms at the side, and the palms of the hands facing forward. Some terms refer to an imaginary midline, or axis, that runs vertically down the center of the body and splits it in two.

- **Medial** means at or nearer to the midline, or on the inner side of; **lateral** means away from the midline, or on the outer side of. For example, the backbone is medial to the kidneys; the left eye is lateral to the bridge of the nose.
- **Superior** means above, or toward the head or upper part of the body; **inferior** means below, or toward the lower part of the body. For example, the superior

vena cava is a large vein that carries blood into the heart from the upper body; the inferior vena cava does the same from the lower body.

- **Anterior** (ventral) means toward the front of the body; **posterior** (dorsal) means toward the back of the body. For example, the heart is anterior to the backbone; the sacrum is posterior to the urinary bladder.
- **Proximal** refers to something that is nearer to the point of attachment of a body part; **distal** means further away. The proximal end of a digit in the hand is at the knuckle, while its distal end is at the fingertip.
- **Superficial** is used to indicate something at or near the body's surface; **deep** means located away from the body's surface. For instance, the skin is superficial to the skeleton, while the brain is deep to the skull.

### Body orientation

Superior

Axis

Lateral

Medial

Distal    Proximal

Anterior

Posterior

Inferior

### Body cavities

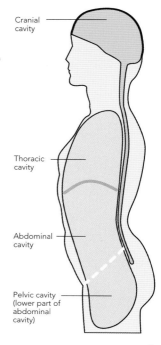

Cranial cavity

Thoracic cavity

Abdominal cavity

Pelvic cavity (lower part of abdominal cavity)

▶ *The cranial cavity contains the brain. The thoracic cavity houses the lungs and heart. The abdominal cavity contains most of the organs of the digestive system, the liver, the kidneys, and the spleen; its lower part, sometimes called the pelvic cavity, contains the reproductive organs and the urinary bladder.*

# CELLS

Cells are the basic units of life. The body contains more than 100 trillion (100,000,000,000,000) cells that work together to form a functional human being. There are more than 200 different types of cells – including red blood cells, muscle fibers, and nerve cells – each with their own particular role. All cells share the same basic structure. A thin, pliable plasma (or cell) membrane surrounds the gelatin-like cytoplasm, which contains various organelles ('little organs'), such as mitochondria and ribosomes, each with their own function; the spherical nucleus is the control center of the cell.

## Plasma membrane
Merely 8 nanometers (0.000008 mm) in diameter, the plasma membrane forms a boundary that separates the cell from its surroundings and selectively controls what substances enter or leave the cell.

Many cell organelles, including mitochondria, are bound by a membrane or membranes with a structure similar to that of the plasma membrane.

## Cytoplasm
The cytoplasm is the part of the cell that lies between the plasma membrane and the nucleus. It consists of a viscous fluid, the cytosol, which is 90 percent water and contains dissolved amino acids, salts, sugars and other substances, and organelles. These include microfilaments (thin protein strands) and microtubules (hollow protein rods) that together form a cytoskeleton which gives the cell its shape and keeps other organelles in their correct positions.

## Organelles
The organelles divide the cytoplasm into many linked compartments thus ensuring

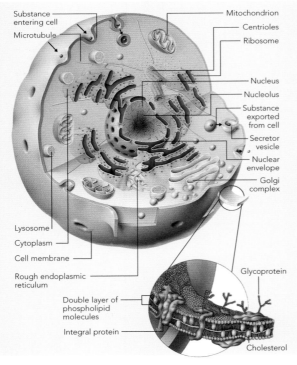

Substance entering cell
Microtubule
Mitochondrion
Centrioles
Ribosome
Nucleus
Nucleolus
Substance exported from cell
Secretor vesicle
Nuclear envelope
Golgi complex
Lysosome
Cytoplasm
Cell membrane
Rough endoplasmic reticulum
Double layer of phospholipid molecules
Integral protein
Glycoprotein
Cholesterol

*Cell and membrane structure*
*This picture illustrates the organelles described in the text above. It also shows the detailed structure of the cell membrane. This consists of a bilayer (double layer) of phospholipid molecules that gives the plasma membrane its flexibility; simple substances, such as oxygen, pass through the bilayer by simple diffusion. It also contains cholesterol, another type of lipid, as well as proteins. Integral proteins pass across the membrane and act as channels to take substances into the cell by a process called active transport, which requires energy. Surface proteins – often attached to sugar molecules to form glycoproteins – act as 'flags' that identify the cell as 'self' to the body's immune system.*

that the many chemical reactions going on inside the cell do not interfere with each other.

- **Mitochondria** (sing. mitochondrion) are sausage-shaped organelles that act as the powerhouses of the cell. Using an oxygen-requiring process, called aerobic respiration, they release energy in the form of ATP (adenosine triphosphate) from glucose and other energy-rich nutrients that is used to drive other chemical reactions in the cell. During the first stage of cell respiration, called glycolysis, which takes place in the cytoplasm, glucose is broken down into a simple molecule called pyruvate. This moves into the mitochondrion (see below/opposite) and is further broken down to release its energy which is transferred to the energy-carrier ATP.

- **Ribosomes** are small granular organelles produced in the nucleolus. They are the site of protein synthesis, a key cellular activity described in more detail on pp. 8–9.

- **Endoplasmic reticulum** (ER) is a system of flattened, interconnected, folded membranes that acts as the cell's 'factory' by manufacturing, storing, and transporting a range of substances. Rough ER is covered with ribosomes; proteins made by ribosomes are either stored here temporarily or are processed for export. Smooth ER lacks ribosomes and is involved in lipid synthesis.

- **Golgi complex** is a set of flattened, membrane sacs that resembles a stack of dinner plates. It processes proteins from rough ER, and packages them in secretory vesicles that pinch off, migrate to the cell membrane, and discharge their contents to the outside. Other vesicles remain in the cell as lysosomes.

- **Lysosomes** are small membrane-bound sacs that are filled with digestive enzymes. Their role is to digest worn-out organelles, and foreign substances so their components can be recycled and reused.

- **Centrioles** are small, columnlike organelles that occur in pairs at right angles to each other and which play a part in cell division.

- **Nucleus** is the cell's control center and is separated from the cytoplasm by a double membrane, the nuclear envelope, which has pores. Its role in cell activities is described more fully on pp. 8–9.

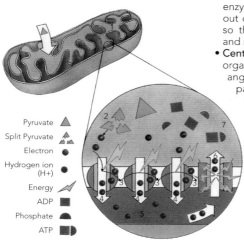

Pyruvate
Split Pyruvate
Electron
Hydrogen ion (H+)
Energy
ADP
Phosphate
ATP

▲ *The Proton Powerhouse*
*Cells are powered primarily by energy released from ATP (adenosine triphosphate) as it becomes ADP (adenosine diphosphate). ATP is made in the mitochondria (1) by recycling ADP. Pyruvate (2)* *is split into carbon dioxide, hydrogen, and high-energy electrons. The electrons pass along a line of proteins in the inner membrane (3) providing energy to pump out protons (4) into the intermembrane space (5). As more protons are* *pumped out, pressure builds up forcing protons back across the membrane. But the protons can only flow back into the matrix via the ATP generator (6) – the enzyme ATP synthetase – and produce ATP (7).*

# HOW CELLS WORK

## Cell chemistry

Inside a cell, a complex sequence of thousands of chemical reactions – collectively known as **metabolism** – takes place every second. Metabolism itself consists of two linked processes: anabolism and catabolism. **Anabolism** consists of those processes, such as protein synthesis, that manufacture substances needed by the cell to build or run itself. **Catabolism** consists of those processes, such as the breakdown of glucose during aerobic respiration, that release energy to power anabolism. Raw materials for both anabolism and catabolism are provided by nutrients absorbed from the small intestine following digestion.

## Enzymes

These protein catalysts accelerate the rate of chemical reactions by thousands or millions of times – thereby making life possible – without being changed or consumed themselves. Each enzyme has a specific three-dimensional shape, and each is specific to a particular chemical reaction. Reactions are catalyzed in part of the enzyme called the 'active site'.

## Chromosomes and DNA

Chromosomes are long, threadlike structures located in the nucleus of every body cell. Most body cells are diploid: that is, they contain two sets of chromosomes, with 23 chromosomes in each set (46 in all); 23 of the chromosomes are inherited from the person's mother (maternal) and 23 from their father (paternal). Chromosomes of the same size – one maternal and one paternal – form matching pairs, called homologous pairs. Sex cells – sperm and ova – are haploid: they each contain just one set of chromosomes. When they fuse during fertilization, the diploid chromosome number is restored.

Chromosomes are formed from deoxyribonucleic acid (DNA), a macromolecule unique in its ability to replicate

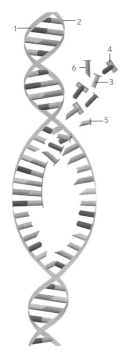

◀ *DNA molecules form a double helix, with two spiral backbones (**1,2**). These are made up of sugar and phosphate units. Linking the backbones are the bases; adenine (**3**), thymine (**4**), guanine (**5**), and cytosine (**6**). Each backbone contributes one base to each rung, which are paired; adenine with thymine, and cytosine with guanine.*

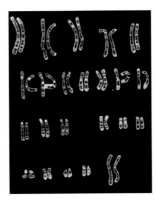

◀ *Human karyotype*
*This karyotype shows a full set of chromosomes from a human body cell. Normally chromosomes are long, thin, and tangled. But just before cell division (see opposite) they become much shorter and thicker, and they replicate, the two duplicate strands being held together centrally by a centromere, forming the characteristic 'X' shape. It is at this stage that these chromosomes have been photographed. The photograph was then cut up, and the chromosomes arranged in homologous pairs. The pairs were then arranged in size order from chromosome pair 1 through to pair 22. The 23rd pair, the sex chromosomes, are either XX (female) or XY (male).*

itself. DNA consists of two intertwined spiraling strands that form a double helix similar in shape to a twisted ladder. The sides of the 'ladder' are formed by a sugar-phosphate backbone; its 'rungs' by pairs of linked bases of which there are four types: adenine (A), cytosine (C), guanine (G), and thymine (T). Within DNA, A only binds with T, and C only binds with G. The DNA molecules inside a set of chromosomes (a genome) have about 35,000 segments, called genes, that each control the production of proteins inside the cell. Each protein has a specified use such as constructing cell components or making enzymes. By controlling protein synthesis, therefore, DNA controls all cell activities and all body features. Within genes, it is the precise but varying sequence of the bases A, C, G, and T that provide the coded instructions used to make proteins. When cells divide (see below) during growth or repair, DNA duplicates itself exactly, so that each new cell has a precise copy of its parent cell's DNA and chromosomes.

## Protein synthesis

Proteins are made up of building blocks called amino acids, of which there are 20 different kinds. The functioning of a protein depends on its shape and this, in turn, depends on the precise order of amino acids in its molecule. During protein synthesis, the cell uses a genetic code to turn the instructions written in bases in the DNA of a gene to a sequence of amino acids in a protein. Each instruction is written in 'words', called codons, made of three 'letters', or bases, such as AAG. The first step, called transcription, is to copy the codons in a gene into a shorter, single stranded molecule called messenger ribonucleic acid (often abbreviated as mRNA). This passes out of the nucleus and into the cytoplasm where the genetic message is translated (see below).

▶ *Translation*
*Once in the cytoplasm, messenger ribonucleic acid (mRNA) attaches itself to a codon-reading machine called a ribosome. As the ribosome progresses along the mRNA, cloverleaf shaped molecules called transfer RNA or tRNA – of which there are many types – deliver their specific amino acid. Which amino acid they carry depends on the sequence of three bases – the anticodon – that projects from their base. The tRNA lines up with the complementary sequence of bases in the codon on mRNA. The amino acids link up through peptide bonds, and the DNA instructions have been successfully translated into an amino acid sequence.*

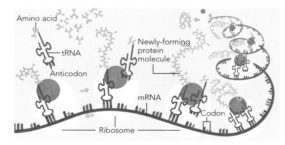

◀ *Mitosis*
*This is the process, used in growth and repair, by which a cell nucleus divides to produce two identical cells. Each chromosome copies itself to form an X-shaped structure composed of two linked identical chromatids. Chromosomes line up in the center of the cell, and then the chromatids are pulled apart to each end of the cell, each forming a new chromosome. After mitosis, the cytoplasm divides to form two new cells.*

Chromosomes shorten and replicate
Chromosomes line up
Chromatids separate
Cell division completed

9

# TISSUE, ORGANS, SYSTEMS

The human body is organized according to a hierarchy of levels of increasing structural complexity from the simplest (molecules), through cells (see pp. 6–7), tissues, organs, and systems, to the most complex, the body itself.

## Tissues

Tissues are aggregations of cells of the same, or similar type, that are grouped together to perform a specific function, and collectively make up the fabric of the body. The study of tissues is known as histology. There are four main categories of tissues: epithelial, connective, muscle, and nervous.

## Epithelial tissue

Also called epithelium (pl. epithelia), epithelial tissues typically form linings inside the body, such as those of the mouth, fallopian tubes, and blood vessels, as well as forming the skin's epidermis. Epithelial tissue consists of a continuous sheet of tightly packed cells, that may be one cell thick or have many cell layers, and which forms a leakproof barrier that prevents the admission of disease-causing microorganisms. It rests on, and is supported by, connective tissue (see below); between them is a basement membrane that reinforces the epithelial sheet by helping it resist tearing and stretching

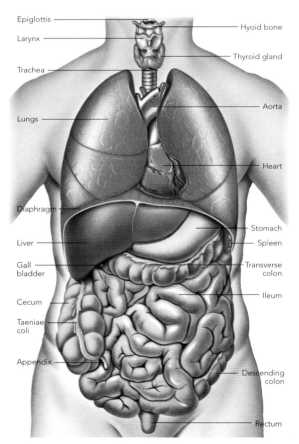

Epiglottis

Larynx

Trachea

Lungs

Diaphragm

Liver

Gall bladder

Cecum

Taeniae coli

Appendix

Hyoid bone

Thyroid gland

Aorta

Heart

Stomach

Spleen

Transverse colon

Ileum

Descending colon

Rectum

◀ *Cell, tissue, organ, system, body*
*This diagram illustrates the hierarchical organization of the body using the digestive system as an example. Four types of cells form the epithelial tissue that lines the stomach and secretes digestive juices. Epithelial tissue is just one of the tissues that makes up the J-shaped organ of the stomach, which is responsible for storing and partially digesting food. The stomach also contains smooth muscle that churns up food; nerve fibers that regulate the rate of muscle contraction; and connective tissue that binds the other layers' tissues together. Stomach, small intestine, and other linked organs form the digestive system, responsible for ingesting, digesting, and absorbing food.*

forces. They are classified in two ways. First, whether they consist of one layer (simple) or many layers (stratified), and second by the shape of their cells – either squamous (flattened); cuboidal (boxlike), or columnar (column-shaped).

## Connective tissue

Connective tissues form the human body's framework by binding, supporting, strengthening, and insulating other tissues. Uniquely, connective tissues consist principally of extracellular material (or matrix) secreted by widely scattered cells. This material has two components: ground substance and protein fibers. Ground substance fills the space between cells. It can be solid, fibrous, gelatinlike, semifluid, or fluid; and contains fibers. Fibers include two main types: collagen fibers are tough and strong; elastin fibers stretch and recoil like rubber bands.

There are four main types of connective tissue. Connective tissue proper supports and binds other tissues. As adipose tissue, it stores fat. Connective tissue is also found in tendons, ligaments, and the skin's dermis. **Cartilage** is a tough but flexible structural tissue that has three forms: hyaline cartilage covers the ends of bones; elastic cartilage is found in the external ear and epiglottis; fibrous cartilage forms the intervertebral disks. Bone, with its hard matrix, supports the body and protects its internal organs. Blood, with its liquid matrix, is the transport medium of the cardiovascular system.

Connective and epithelial tissues combine to form the membranes that line body cavities and hollow organs. Cutaneous membrane forms the skin. Moist mucous membrane (mucosa) lines body cavities, such as the digestive system, that open to the outside. Serous membranes (serosa) covers and lubricates the walls of closed body cavities, such as the thoracic cavity, and the organs inside them; and allows them to slide painlessly over each other.

## Muscle tissue

Muscle tissue consists of well-vascularized cells, called fibers, that can contract, thereby causing movement. There are

three types of muscle tissue (see pp. 28–29): **skeletal muscle** moves, and maintains the posture of the body; **smooth muscle** in the walls of hollow organs moves materials through the body; and **cardiac muscle**, found only in the heart, pumps blood along the blood vessels.

## Nervous tissue

Found only in the nervous system (see pp. 34–35), nervous tissue controls and regulates body functions. It consists of two types of cells. **Neurons** generate and conduct nerve impulses at high speed. **Neuroglia**, or glial cells, support, protect, and insulate neurons.

## Organs and systems

An organ – such as the eyes, heart, or skin – is a structure that carries out a particular function or functions, and which is made up of two or more types of tissues. Linked organs – such as the brain, spinal cord, nerves, and sense organs – form one of the 12 systems in the body, in this case the nervous system.

## Systems

The body's 12 systems interact to form a complete, functioning human being. The integumentary system forms the body's external covering. The skeletal system supports the body, protects its organs, and provides anchorage for skeletal muscles. The muscular system moves the body and maintains posture. The nervous system controls and regulates the body. The endocrine system secretes hormones that regulate processes such as growth and reproduction. The cardiovascular system transports nutrients, oxygen, and other materials to body cells. The lymphatic system drains excess fluid from the tissues. The immune system defends the body against infection, by utilizing white blood cells and chemicals in the blood, lymphatic system, and tissues. The respiratory system takes oxygen into the body and removes carbon dioxide. The digestive system supplies the body with nutrients. The urinary system eliminates metabolic waste and regulates water balance. The reproductive systems (male and female) produce offspring.

## Skin

Skin is the largest body organ. It weighs up to 11 pounds (5 kg) and covers an area of about 21.5 sq ft (2 sq m), forming an interface between the body's interior and its surroundings. As a physical barrier, skin stops water from leaking out of or into the tissues; prevents the entry of bacteria and other disease-causing microorganisms; filters out the harmful and potentially carcinogenic ultraviolet (UV) radiation in sunlight; and repairs itself if cut or torn. Skin also helps maintain the body's temperature at a constant 98.6°F (37°C); and contains a range of sensory receptors (see pp. 46–47).

## Skin structure

Skin has two layers: the epidermis and dermis. The **epidermis** is the thin, upper part of the skin. Its main role is protective. Its uppermost layer consists of dead, flattened cells, packed with keratin (a tough, waterproof protein), that are constantly worn away by everyday wear and tear, releasing scaly skin flakes; they are replaced by living cells in deeper layers that divide continuously. Other epidermal cells, called melanocytes, produce the brown pigment melanin, which gives skin its color and forms a protective screen which absorbs harmful UV rays. People with darker skin produce more melanin; prolonged exposure to sunlight increases melanin production and temporarily darkens the skin.

The **dermis** is the lower, thicker part of the skin. A network of collagen and elastin fibers gives this connective tissue, respectively, strength and elasticity. The dermis contains blood vessels that play a part in temperature regulation (see opposite); sensory nerve endings and receptors that detect external stimuli; deep pits called hair follicles from which hairs grow; and sebaceous glands that release an oily, bactericidal liquid called sebum that helps keep the skin and hairs soft, flexible, and waterproof. Sweat glands release watery sweat onto the surface of the skin in warm conditions, from where it evaporates to

### Section through the skin

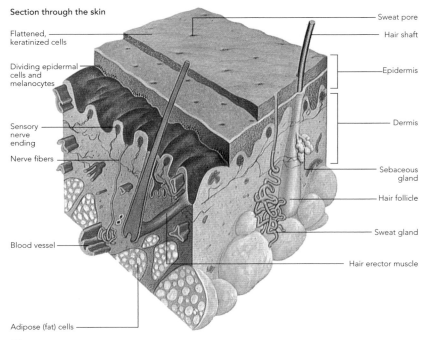

Flattened, keratinized cells

Dividing epidermal cells and melanocytes

Sensory nerve ending

Nerve fibers

Blood vessel

Adipose (fat) cells

Sweat pore

Hair shaft

Epidermis

Dermis

Sebaceous gland

Hair follicle

Sweat gland

Hair erector muscle

cool the body. Adipose (fat) cells beneath the dermis help insulate the body.

## Hair

Except for the palms, soles, lips, nipples, and parts of the external genitals, millions of hairs are distributed over the surface of the skin, with more than 100,000 on the scalp alone. Human hair has no insulatory role except on the scalp, where it also protects the skin from sunlight. Hair acts as a sensor, and keeps dust out of the eyes and nose. Hairs are tubes of dead cells filled with keratin that grow upward and above the surface of the skin from living, dividing cells in the base of hair follicles. Scalp hair grows at about 0.08 in (2 mm) a week for three to five years, and is then pushed out by a newly-growing hair. In men over 30, the male sex hormone testosterone can disrupt this cycle and cause male pattern baldness. Hair color is dependent on melanin content; hair becomes white with age due to lack of pigmentation. There are two types of body hair. Fine vellus hairs cover the bodies of children and women. Coarser terminal hairs are found in the scalp and eyebrows of all humans, in the armpits and pubic regions of adults, and in the body and facial hair of adult males.

## Nails

Nails are scaly extensions of the epidermis that cover and protect the upper, distal ends of the fingers and toes, and aid gripping. A nail is made of dead cells packed with keratin, and grows from living cells at its base at about 0.2 in (5 mm) per month (slower in toenails). Each nail consists of a free edge that projects from the tip of the finger or toe; a body, covering the epidermis, and colored pink by the underlying dermis; and a root, embedded in the skin.

Proximal skin fold or cuticle
Lunula
Body of nail
Nail bed
Free edge of nail
Nail root
Nail matrix
Distal phalanx of finger

▲ **Section through nail**
*The nail bed underlies the nail and consists of the deeper, living layer of the epidermis. At the proximal end of the nail bed is the thicker nail matrix. Nail growth happens as living cells in the nail matrix divide, push forward, and become keratinized to form the nail body, which slides over the nail bed. Much of the nail body appears pink, but a white crescent or lunula covers the nail matrix. Skin folds, called nail folds, overlap the proximal and lateral edges of the nails.*

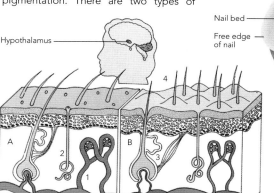

Hypothalamus

▲ **Temperature control**
*The temperature center in the hypothalamus controls heat loss and heat production by the body through the skin (below). Overheating (A) produces an increased blood flow from the blood vessels (1), which radiates heat and causes sweating through*

*the sweat glands (2), thereby losing heat. A fall in temperature (B) constricts the surface blood vessels, stops sweating and makes the erector muscles (3) contract, causing the hairs (4) to stand on end, trapping air as an insulating layer. Additional heat can be produced by shivering.*

# THE SKELETON

The skeleton provides a strong, internal framework that supports the body, makes up about 20 percent of its weight, and consists of 206 bones. These bones meet at joints, the majority of which are freely movable, making the skeleton flexible and mobile. The skeleton also contains cartilage and ligaments. **Cartilage** is a tough, flexible connective tissue that forms the frameworks of the ear and nose, links the ribs to the sternum, and covers the ends of bones inside joints. **Ligaments** are strong strips of fibrous connective tissue that hold bones together at joints, thereby stabilizing the skeleton during movement. The bones of the skeleton are divided into two broad groups: the axial and appendicular skeletons.

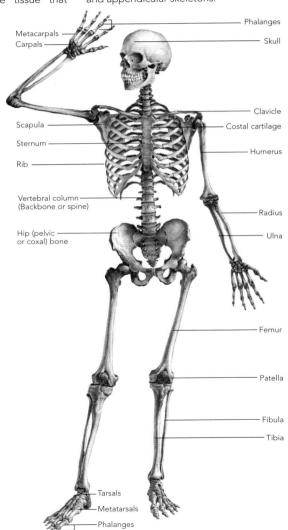

Metacarpals
Carpals
Phalanges
Skull
Scapula
Clavicle
Costal cartilage
Sternum
Humerus
Rib
Vertebral column (Backbone or spine)
Hip (pelvic or coxal) bone
Radius
Ulna
Femur
Patella
Fibula
Tibia
Tarsals
Metatarsals
Phalanges

▶ *Axial skeleton*
*The axial skeleton consists of 80 bones that form the body's long axis and are grouped into three regions: skull, vertebral column, and rib cage. The skull is made up of 22 bones and contains six ossicles, or ear bones; the hyoid bone is found in the neck. The vertebral column consists of 26 irregular bones. The rib cage consists of 12 pairs of curved ribs linked to the sternum by flexible strips of costal cartilage. The axial skeleton serves to support the head, neck, and trunk; and protects the brain, spinal cord, lungs, and heart.*

## Other skeletal functions

In addition to supporting the body, the skeleton has other important functions. It surrounds and protects soft internal organs, such as the brain, heart, and lungs, from external knocks and blows. It provides anchorage points for the tendons that link bones to skeletal muscles, which pull on bones across joints and move the body. Bones store the mineral calcium, needed for normal nerve and muscle function, releasing it into, or removing it from, the bloodstream according to the body's demands. Red bone marrow inside certain bones manufactures red and white blood cells. Yellow bone marrow inside long bones, such as the femur, stores fat.

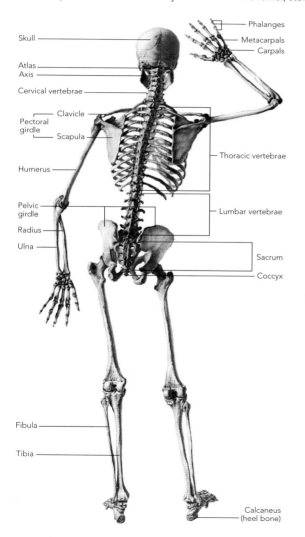

Skull

Atlas
Axis

Cervical vertebrae

Clavicle
Pectoral
girdle
Scapula

Humerus

Pelvic
girdle

Radius

Ulna

Phalanges
Metacarpals
Carpals

Thoracic vertebrae

Lumbar vertebrae

Sacrum

Coccyx

Fibula

Tibia

Calcaneus
(heel bone)

◀ *Appendicular skeleton*
*The 126 bones of the appendicular skeleton make up the upper and lower extremities (limbs) and the girdles that attach them the axial skeleton. Both upper and lower limbs consist of three sections. The upper limb consists of 27 hand bones; the radius and ulna (forearm bones); and the humerus (arm bone). The humerus articulates with a weak but very mobile pectoral girdle, consisting of the scapulas and clavicles. The lower limb consists of 26 foot bones; the tibia and fibula (leg bones); and the femur (thigh bone) and patella (kneecap). The femur articulates with a strong, almost rigid pelvic girdle made up of two hip bones.*

# BONE

ar from being dry and dusty, living bones are organs that are moist, have a rich supply of blood and lymph vessels and nerves, and are constantly being reshaped. Bones are made of a hard connective tissue that consists of a matrix containing widely scattered bone cells called osteocytes. The matrix has two components. About 65 percent consists of mineral salts, especially calcium phosphate, which gives the bone hardness. Wrapped around these crystals is the fibrous protein collagen, which makes up about 35 percent of the matrix and gives some flexibility to bone. Together these elements make bone as strong as steel, but much lighter, and gives it a degree of springiness that enables it to withstand the stresses generated by supporting and moving the body without fracturing.

## Inside bones

All bones are surrounded by periosteum, an outer protective 'skin' made of fibrous connective tissues through which pass the blood vessels that supply blood cells. Periosteum also contains cells that remodel bones in response to the stresses they experience: osteoclasts break cells down, while osteoblasts build them up. Within the periosteum is a layer of compact, or cortical, tissue second only in hardness to tooth enamel. Compact bone consists of macroscopic cylinders of bone matrix called osteons that run in parallel, each acting as a weight-bearing

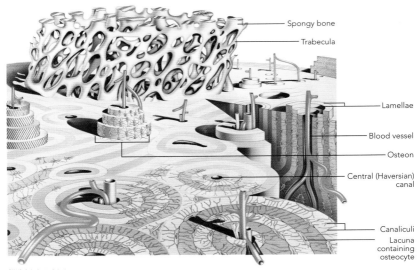

Spongy bone

Trabecula

Lamellae

Blood vessel

Osteon

Central (Haversian) canal

Canaliculi

Lacuna containing osteocyte

▲ *Compact and spongy bone*
*This section though a bone shows the outer compact bone and the inner spongy bone. The cylindrical osteons that make up compact bone consist of concentric tubes (lamellae) of matrix surrounding a central (Haversian) canal that carries blood vessels.*

*These supply widely spaced bone cells, called osteocytes, that are 'trapped' within spaces, called lacunae, found between lamellae. Osteocytes communicate with each other through tiny 'tentacles' that pass along hairlike canals called canaliculi. Spongy bone consists of struts, called*

*trabeculae, made of irregularly arranged lamellae and osteocytes, and spaces. Trabeculae are arranged along lines of stress, an arrangement that gives spongy bone considerable strength. The spaces make spongy bone much lighter than the compact bone that surrounds it.*

pillar and collectively conferring great strength. Within compact bone is honey-comblike spongy, or cancellous, bone. Despite its structure, spongy bone retains considerable strength, but its lightness reduces the overall weight of the skeleton. The spaces within spongy bone – and the central medullary cavity in long bones such as the femur – are filled with soft bone-marrow. At birth, all bone marrow is red, the type that produces blood cells. By adulthood, red bone marrow in the long bones has been replaced by yellow marrow which provides a fat store.

## Bone growth

Inside the uterus, the skeleton of the early fetus is made of hyaline cartilage. Gradually osteoblasts replace this cartilage framework with bone matrix. This process, known as ossification, continues from birth to adolescence. Osteoblasts develop growing bones from the outside and then become embedded within the matrix and change roles to become mature bone cells or osteocytes. In adulthood, although growth ceases, bones are constantly renewed, or remodeled, to strengthen those parts that are subjected to greatest stresses. Remodeling is achieved by the balanced activities of osteoclasts, which erode bone tissues, and osteoblasts which rebuild it. Both osteoclasts and osteoblasts play a role in regulating levels of calcium in the blood.

## Bone repair

Despite their strength, bones may break, or fracture, either as a result of extreme pressure caused by a fall or other accident, or because of diseases such as osteoporosis. In a simple (closed) fracture, the broken ends of bones remain under the skin. In a compound (open), fracture, broken ends project out though the skin. As living tissue, bones have their own built-in repair system. Where possible, fractured bones need to be immobilized by a plaster cast or, in more severe cases, by pinning, to ensure bones are correctly aligned as they heal.

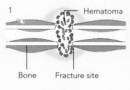

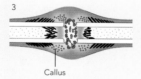
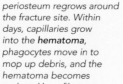

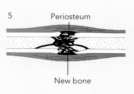

*Bone repair*
A fractured bone goes through several stages of repair before it is fully healed. Around six hours after the injury, a large hematoma (or blood clot) forms at the fracture site. This seals off any ruptured blood vessels and the site becomes inflamed. The periosteum regrows around the fracture site. Within days, capillaries grow into the **hematoma**, phagocytes move in to mop up debris, and the hematoma becomes replaced by a fibrous cartilage callus in which osteoblasts start to lay down new bone and hold two broken ends together. About four weeks after fracture, a bony callus of spongy bone forms. In the next few months, the bone is remodeled by osteoclasts to remove excess material on the bone's shaft, and an outer layer of compact bone is laid down to complete the repair.

# THE SKULL

The human skull shapes the head and face, protects the fragile brain, and houses and protects the special sense organs for taste, smell, hearing, vision, and balance. It is constructed from 22 bones, 21 of which are locked together by immovable joints, known as **sutures**, to form a structure of great strength. Blood vessels and cranial nerves enter and leave the skull through holes called **foramina** (sing. foramen) and canals. Skull bones divide into two groups: cranial bones and facial bones.

## Cranial bones

The eight bones of the **cranium** support, surround and protect the brain within the cranial cavity. They form the roof, sides, and back of the cranium, as well as the cranial floor on which the brain rests. The frontal bone forms the forehead, the anterior part of the cranial floor, and the roof of the orbits (eye sockets). Two parietal bones form the roof and sides of the cranium. Two temporal bones form the inferior lateral parts of the cranium, and part of the cranial floor. An opening in the temporal bone, the **external auditory meatus**, directs sounds into the inner part of the ear that is encased within, and which contains three small, linked bones called ossicles. The occipital bone forms the posterior part of the cranium and much of the cranial floor. The **occipital bone** has a large opening, the **foramen magnum**, through which the brain connects to the spinal cord. The **occipital condyles** articulate with the atlas (first cervical vertebra), enabling nodding movements of the head. The **ethmoid bone** forms part of the cranial floor, the medial walls of the orbits, and the upper part of the nasal septum, which divides

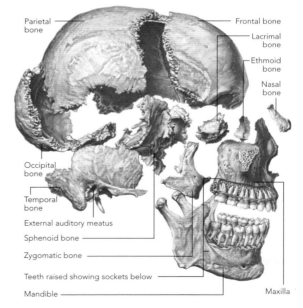

*Parietal bone*

*Frontal bone*

*Lacrimal bone*

*Ethmoid bone*

*Nasal bone*

*Occipital bone*

*Temporal bone*

*External auditory meatus*

*Sphenoid bone*

*Zygomatic bone*

*Teeth raised showing sockets below*

*Mandible*

*Maxilla*

▶ *Skull bones*
*This exploded view of the skull shows many of the component bones that make up its structure. The exposed edges of most bones, apart from the mandible, can be seen to have jagged or serrated edges. These edges fit together like pieces of a jigsaw to form immovable joints called sutures that are found only in the skull and which give it enormous strength and rigidity. The inner surface of the cranial bones is attached to the meninges (membranes) which surround, protect, and stabilize the brain. Depressions in the inner* *cranial surface accommodate blood vessels that supply the meninges. The outer surfaces of both cranial* *and facial bones provide areas of attachment for muscles that move the head and produce facial expressions.*

the nasal cavity vertically into left and right sides. The **sphenoid bone**, which is shaped like a bat's wings, acts as a keystone by articulating with, and holding together, all the other cranial bones.

## Facial bones

The 14 facial bones form the framework of the face; provide cavities for the sense organs of smell, taste, and vison; anchor the teeth; form openings for the passage of food, water, and air; and provide attachment points for the muscles that produce facial expressions. Two **maxillae** form the upper jaw, contain sockets for the 16 upper teeth, and link all other facial bones apart from the mandible (lower jaw). Two **zygomatic bones**, or cheekbones, form the prominences of the cheeks and part of the lateral margins of the orbits. Two **lacrimal bones** form part of the medial wall of each orbit. Two **nasal bones** form the bridge of the nose. Two **palatine bones** form the posterior side walls of the nasal cavity and the posterior part of the hard palate. Two inferior **nasal conchae** form part of the lateral wall of the nasal cavity. The **vomer** forms part of the nasal septum. The **mandible**, the only skull bone that is able to move, articulates with the temporal bone allowing the mouth to open and close, and provides anchorage for the 16 lower teeth.

## Sinuses

Sinuses are air-filled spaces found in the frontal, sphenoid, ethmoid, and paired maxillae, clustered around the nasal cavity. These spaces reduce the overall weight of the skull.

## Skull development

In the fetus, skull bones are formed by intramembranous ossification. A fibrous membrane ossifies to form skull bones linked by areas of as yet unossified areas of membrane called **fontanelles**. At birth, these flexible areas allow the head to be slightly compressed, and permit brain growth during early infancy.

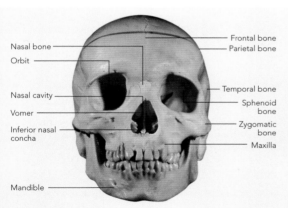

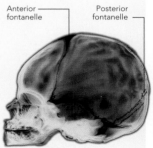

▲ **Neonate skull**
This skull of a newborn child shows two of the prominent fontanelles present at birth: the anterior fontanelle between the parietal and frontal bones, and the posterior fontanelle between the parietal and occipital bones. Fontanelles are replaced by bone between one and two years after birth.

▲ **Front view of skull**
Each orbit is a bony cavity formed from parts of seven interlocking bones: frontal, zygomatic, sphenoid, maxilla, ethmoid, lacrimal, and palatine. It encases and protects the eyeball, and provides attachment points for the muscles that move the eye. The nasal cavity is formed from parts of the ethmoid, palatine, maxillae, vomer, inferior nasal conchae, and by cartilage. It warms and moistens inhaled air, and houses olfactory (smell) receptors.

# BACKBONE AND RIBS

Together with the sternum and ribs, the backbone (also known as the vertebral column, spinal column, or spine) forms the skeleton of the trunk. The backbone consists of a chain of irregular bones called **vertebrae** that meet at slightly movable joints. Each joint permits only limited movement, but collectively the joints give the backbone considerable flexibility enabling it to rotate, and to bend anteriorly, posteriorly, and laterally. The average backbone makes up about 40 percent of body height. It extends from the skull to its anchorage in the pelvic girdle, where it transmits the weight of the head and trunk to the lower limbs. It also supports the skull; encloses and protects the delicate spinal cord; and provides an attachment point for the ribs, and for the muscles and ligaments that support the trunk of the body.

## Intervertebral disks

Intervertebral disks are found between adjacent vertebrae from the second cervical vertebra (axis) to the sacrum. Each disk has an inner soft, pulp nucleus covered by an outer fibrous covering of fibrous cartilage. Each disk forms a strong, slightly movable joint. Collectively, disks cushion vertebrae against vertical shocks, and allow various movements of the backbone.

## Vertebral curves

Seen in side view (left) a normal backbone has four curves that give it an S-shape. The cervical and lumbar curves are convex anteriorly, while the thoracic and sacral curves are concave anteriorly. The S-shape allows the backbone to function as a spring rather than a flexible rod, thereby absorbing shock during walking and running; enhancing the strength and flexibility of the backbone; and facilitating balance when upright by placing the trunk directly over the feet.

## Bony thorax

The cone-shaped bony thorax surrounds the thoracic cavity, and is formed by 12

Atlas
Axis
Cervical vertebrae
1
1 Cervical curve
2 Thoracic curve
3 Lumbar curve
4 Sacral curve
2
Thoracic vertebrae
Intervertebral disk
Lumbar vertebrae
3
Sacrum
4
Coccyx

◄ *Regions of the backbone*
*An adult backbone consists of 26 vertebrae of which two, the sacrum and coccyx, are composites consisting of vertebrae that fused during childhood. The backbone has five sections. Seven small cervical vertebrae form the neck, the most flexible part of the backbone. The uppermost cervical vertebra, the atlas, articulates with the occipital condyle of the skull to enable nodding 'Yes' movements of the head;*

*articulation of the atlas with the axis, the second cervical vertebra, produces shaking 'No' movement of the head. Twelve thoracic vertebrae each articulate with a pair of ribs. Five large lumbar vertebrae form the hollow small of the back and bear most of the weight of the head and trunk. The triangular sacrum, made of five fused bones, forms a strong anchorage for the pelvic girdle, with which it forms the pelvis. The coccyx, or tailbone, consists of four fused vertebrae.*

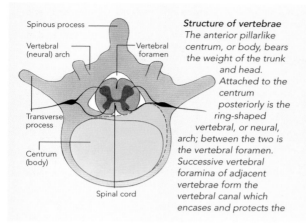

Spinous process

Vertebral (neural) arch

Vertebral foramen

Transverse process

Centrum (body)

Spinal cord

**Structure of vertebrae**
The anterior pillarlike centrum, or body, bears the weight of the trunk and head. Attached to the centrum posteriorly is the ring-shaped vertebral, or neural, arch; between the two is the vertebral foramen. Successive vertebral foramina of adjacent vertebrae form the vertebral canal which encases and protects the spinal cord. Projecting from the vertebral arch posteriorly is a spinous process, and laterally two transverse processes. These provide attachment points for the muscles and ligaments that stabilize the backbone. Paired superior and inferior articular processes bear smooth joint surfaces, called facets, that form movable joints with adjacent vertebrae, in addition to the joints formed by intervertebral disks.

thoracic vertebrae posteriorly, 12 pairs of ribs laterally, and the sternum and costal cartilages anteriorly. Its cagelike structure protects the thoracic and upper abdominal organs, supports the pectoral girdles and upper limbs, and plays a part in breathing.

## Ribs

The ribs are curved, flat bones with a slightly twisted shaft. The 12 pairs of ribs form a rib cage that protects the heart, lungs, major blood vessels, stomach, and liver. At its posterior end, the head of each rib articulates with the facets on the centra of adjacent vertebrae, and with a facet on a transverse process. These vertebrocostal joints are plane joints that allow gliding movements. At their anterior ends, the upper ten pairs of ribs attach directly or indirectly to the sternum by flexible costal cartilages. Together, vertebrocostal joints and costal cartilages give the rib cage sufficient flexibility to make movements up and down during breathing. Ribs 11 and 12 are 'floating' ribs that articulate with the sternum indirectly via the costal cartilage of another rib or not at all.

▶ *Rib cage*
*The 12 pairs of ribs in the rib cage increase in size up to the seventh pair, and then they decrease. The seven superior pairs of ribs, known as the true or vertebrosternal ribs, attach directly to the sternum through their own costal cartilages. The remaining five pairs of ribs are called false ribs. The costal cartilages of the eighth to tenth pairs, the vertebrochondral ribs, attach to each other and to the costal cartilage of the*

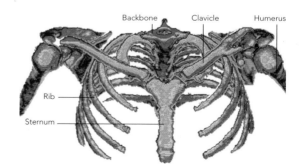

Backbone    Clavicle    Humerus

Rib

Sternum

*seventh pair of ribs. These costal cartilages form the inferior, or costal, margin of the rib cage. The eleventh*

*and twelfth ribs, the floating or vertebral ribs, have no anterior attachment to the sternum.*

# THE SHOULDER AND ARM

Unlike their close relatives in the animal kingdom, humans have an upright stance and walk on their hind limbs or lower limbs. This leaves their forelimbs – the upper limbs – free to pick up and manipulate objects. The bones of the shoulder and arm are, accordingly, thinner and more mobile than the stronger weight-bearing bones of the lower limbs.

## Pectoral girdle

The pectoral, or shoulder, girdle is formed on each side of the body by two bones: the **clavicle**, or collarbone, and the **scapula**, or shoulder blade. Together, these two bones and their associated muscles form the shoulder. The pectoral girdle attaches the upper limbs to the trunk, and provides anchorage points for the muscle that move the upper limbs. Unlike the pelvic (hip) girdle, the pectoral girdle is light and highly mobile, and it is

not attached to the backbone. The long, curved clavicle articulates with the sternum at its medial end and scapula at its lateral end. It acts as strut which holds the scapula and arm away from the upper part of the thorax. The triangular scapula lies on the dorsal surface of the rib cage. Its lateral margin, next to the armpit, forms a shallow, cuplike cavity, the glenoid cavity. The head of the humerus fits into the glenoid cavity to form a ball and socket joint, the shoulder joint. This joint is highly mobile, but not very stable, because the glenoid cavity is shallow and poorly reinforced and does not restrict the movement of the humerus.

## Upper limb

The upper limb consists of 30 bones organized into three sections: the arm, forearm, and hand. The **humerus**, the single bone of the arm and the largest

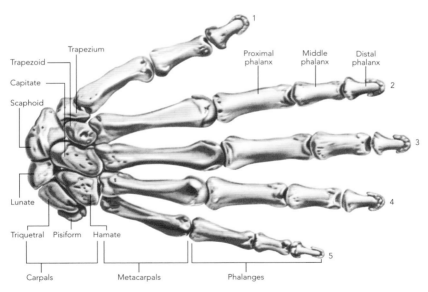

| Trapezium | | | Proximal phalanx | Middle phalanx | Distal phalanx |

Trapezoid
Capitate
Scaphoid

Lunate
Triquetral · Pisiform · Hamate

Carpals · Metacarpals · Phalanges

1
2
3
4
5

▲ *The hand*
*Hand bones fall into three groups. Arranged in two rows, eight short carpal bones form the wrist, which forms a flexible bridge*

*between forearm and palm. Five metacarpal bones radiate from the wrist to form the palm of the hand. Their distal ends form the knuckles. Fourteen long bones called*

*phalanges (sing. phalanx) form the digits, or fingers. Each finger has three phalanges – proximal, middle, and distal – while the pollex, or thumb, has two.*

and longest in the upper limb, articulates proximally with the scapula at the shoulder joint, and distally in a hinge joint with the forearm bones, the **ulna** and **radius**, at the elbow. Running parallel in the forearm, the radius and ulna articulate distally with **carpals** in the wrist to form the wrist joint. At their proximal end they articulate with each other at a pivot joint, which allows the radius to cross over the ulna thereby enabling the hand to rotate through 180°. Each hand consists of 27 small bones that form the most flexible and versatile part of the skeleton which is strong enough to grip objects firmly, precise enough to write a letter, yet delicate enough to thread a needle.

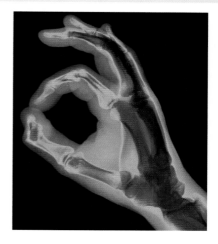

▲ *The opposable thumb*
*A highly flexible saddle joint between the metacarpal of the thumb and the trapezium (a carpal) makes the thumb the most mobile of the digits. The thumb can swing across the palm and touch the tips of the other fingers in turn, an action called opposition. The opposable thumb enables the hand to grip, including the precise grip shown in this X-ray.*

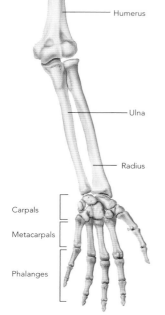

Humerus

Ulna

Radius

Carpals

Metacarpals

Phalanges

Clavicle

Scapula

Glenoid cavity

Humerus

▲ *Shoulder joint*
*Clearly visible here are the bones that form the shoulder joint. The ball-like upper end of the humerus fits into the shallow glenoid cavity of the scapula to form a highly mobile but unstable joint.*

▲ *Shoulder and arm*
*The arrangement of bones in the shoulder, arm and hand provide great flexibility and a wide range of movement. The arm can* *be extended forward to grasp objects, then folded back against the body. It can also move out to the side, upward, backward, and swing around.*

# HIP AND LEG

## Pelvic girdle
The pelvic, or hip, girdle is the strong, rigid structure which attaches the legs to the trunk, and transmits the weight of the upper body to the legs. It consists of two **coxal**, or hip, bones which meet anteriorly at the pubic symphysis, and posteriorly are firmly connected to the **sacrum**.

## Hip bones
Each hip bone is made up of three fused bones: the ilium, ischium, and pubis. The **ilium** forms the largest part of the hip bone. Posteriorly, each ilium articulates with the sacrum at the sacroiliac joint through which the weight of the upper body is transmitted from the backbone to the pelvic girdle. The L-shaped **ischium**, which is inferior and posterior, bears a person's weight when they sit. The **pubis**, or pubic bone, is anterior and is linked to its partner by a fibrous cartilage disk to form a slightly movable joint. Where the three bones meet on the lateral surface of each hip bone, they form a deep, cuplike socket called the **acetabulum** into which fits the ball-like head of the **femur** (thigh bone) at

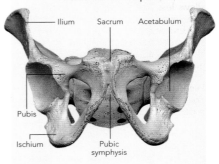

**Anterior view of the pelvis**

*Ilium  Sacrum  Acetabulum*
*Pubis*
*Ischium  Pubic symphysis*

### Hip bones
*This view of the pelvis – shown as if lying down – shows the two fused hip bones that make up the pelvic girdle. The hip bone has three component bones, the ilium, ischium, and pubis. Separate during childhood, the three bones fuse during adolescence. Also seen clearly here is the acetabulum, or socket, that houses the head of the femur in the hip joint.*

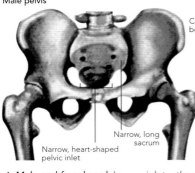

**Male pelvis**

*Narrow, long sacrum*
*Narrow, heart-shaped pelvic inlet*

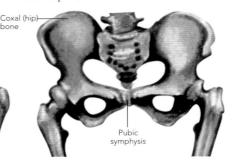

**Female pelvis**

*Coxal (hip) bone*
*Pubic symphysis*

▲ **Male and female pelvis**
*The male and female pelvis show clear differences. The male pelvis is made of heavier, thicker bones; and is narrow and deep. It has a narrow, long sacrum; a narrow pubic arch, the inverted V-shaped arch that is inferior to the pubic symphysis; and a pelvic*

*inlet – the central opening of the pelvis – that is narrow and heart-shaped. The male pelvis is adapted to support the male's heavier build and stronger muscles. The female pelvis is broader and shallower than the male pelvis. It has a wide, short sacrum; a broad pubic arch; and a*

*pelvic inlet that is wide and round. This structure reflects the female's role in childbearing. The broader pelvis supports the fetus during its development. The pelvic inlet is wide enough to allow the fetus's relatively large head to pass through it during childbirth.*

the hip joint. This ball-and-socket joint allows less freedom of movement than the one found in the shoulder between the humerus and pectoral girdle because of the strong ligaments that constrain it. However, although less mobile, the hip joint is much more stable than the shoulder joint.

## Lower limb

The lower limb consists of thirty bones organized into three sections: the thigh,

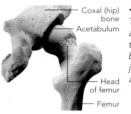

Coxal (hip) bone
Acetabulum
Head of femur
Femur

◀ *Hip and femur*
*The femur and acetabulum fit together forming a ball-and-socket joint that allows for articulation.*

▼ *Hip and leg*
*Linked to the rest of the body through the pelvic girdle, the lower limbs support the entire weight of*

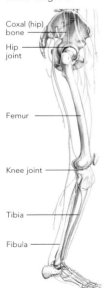

Coxal (hip) bone
Hip joint
Femur
Knee joint
Tibia
Fibula

*the body and withstand the considerable forces generated during walking, running, or jumping. These roles are reflected in the fact that the lower limb bones are much stronger and thicker than those that make up the upper limb.*

leg, and foot. The thigh bone, or **femur**, is the body's longest and strongest bone and bears its weight. At its proximal end, its rounded head forms the hip joint with the acetabulum of the pelvic girdle. Where its laterally angled neck joins the long shaft, there are projections called **trochanters** to which thigh and buttock muscles attach. At its distal end the femur widens into two condyle to form the knee joint with the proximal end of the tibia. The triangular **patella**, or kneecap, protects the knee joint anteriorly. The leg has two parallel bones, the tibia and fibula. The **tibia** transmits the body's weight from femur to foot. At its distal end it forms a hinge joint – the ankle joint – with the talus, one of the tarsal bones in the foot. The slender **fibula** articulates at both ends with the tibia and adds stability to the leg. The broad feet bear the body's weight, help keep it balanced during movement and when stationary, and act as levers that push the body forward during movement.

## Pelvis

Together, the pelvic girdle and sacrum form the pelvis. Reinforced by strong ligaments, the basin-shaped pelvis supports the abdominal organs; and surrounds and protects the organs of the lower abdomen including the reproductive organs and the urinary bladder. It also provides anchorage for powerful muscles that move the legs and support the trunk.

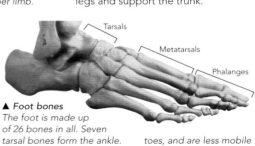

Tarsals
Metatarsals
Phalanges

▲ *Foot bones*
*The foot is made up of 26 bones in all. Seven tarsal bones form the ankle. Five metatarsals form the sole of the foot which, reinforced by ligaments and tendons, forms a springlike arch that absorbs impact during walking and running. The 14 phalanges form the*
*toes, and are less mobile than those in the fingers. The arrangement of the bones in the foot allow little free movement compared to the hand but enable the foot to act as a weight bearer and mover.*

# JOINS

A joint, or articulation, is a part of the skeleton where two or more bones meet. Joints allow the skeleton to move. They also hold the skeleton together and give it stability because the bones in a joint are held together by strips of tough fibrous connective tissue called ligaments. In all joints, the ends of bones do not touch directly but are separated by layers of cartilage or fibrous tissue. Joints can be classified into three types – fixed, slightly movable, and movable joints – according to the amount of movement they permit. The more movable a joint is, the less strength it has.

### Fixed joints
Fixed (immovable) joints permit no movement. The most obvious example of this type of joint are the sutures that, like seams, connect the bones of the skull. In a suture, the serrated edges of skull bones lock together like pieces of a jigsaw, reinforced by fibrous tissue between them. This arrangement gives the skull great strength, vital for its role of protecting the brain.

### Slightly movable joints
This type of joint allows a small amount of movement, and provides strength with flexibility. Typically, the articular surface of the bone is covered with hyaline cartilage which is fused to a pad of compressible fibrous cartilage which separates the two bones and permits limited movement. In the backbone, for example, slightly movable joints between adjacent vertebrae allow limited movement, but collectively all the movement of all vertebrae give the backbone considerable flexibility. A further example of this type of joint is the pubic symphysis in the anterior part of the pelvis.

### Movable joints
Movable, or synovial, joints allow bones to move freely, make up most of the joints in the skeleton, and enable the body to perform a wide range of movements including walking, chewing, and writing. While there are different types of movable joints, each with its own range of movement, all have the same basic structure which allows free movement yet also provides stability. Inside the joint, opposing bone surfaces are covered by glassy hyaline (articular) cartilage, a hard-wearing coat that reduces friction between bones when they move and helps to absorb shocks during movement, thereby pre-

*Types of movable human joints*
*These illustrations show the range of movements permitted by six main types of movable joints in the body. In most the bones hinge on each other or rotate. In the plane joint the bones glide over each other.*

**Saddle joint**
Two U-shaped surfaces allow movement in two directions and in a circle. Found at the base of the thumb.

**Condyloid joint**
Allows movement in two directions and in a circle, e.g. knuckle joint

**Hinge joint**
Allows movement in one plane, e.g. flexion and extension of knee or elbow

**Ball and socket joint**
Allows movement in many directions, e.g. shoulder and hip joints

**Plane joint**
Allows bones to make small sliding movements from side to side e.g. tarsal bones in foot

**Pivot joint**
Allows rotation, but no other movement, e.g. between the atlas and axis

venting bone ends from being crushed. Between the cartilage-covered bone ends there is a space called the synovial, or joint, cavity. This is enclosed by the articular, or joint, capsule. Its outer layer is a tough fibrous coat that strengthens the joint by holding the bones together. Its inner layer is the synovial membrane that secretes an oily, yellow liquid, called synovial fluid, into the synovial cavity. Synovial fluid reduces friction further by making the hyaline cartilage more slippery, thereby increasing ease of movement. Movable joints are reinforced by ligaments external to the joint capsule; some, including the knee, have internal ligaments to provide additional reinforcement. The muscles that pass across a joint and link bones on either side of it also play a part in stabilizing it. Some movable joints, such as the knee, have wedges of fibrous cartilage called articular disks, or menisci (sing. meniscus), separating articular surfaces. They improve the fit between bones ends, thereby both reducing wear and tear and stabilizing the joint.

Movable joints are divided into six types – plane, hinge, pivot, condyloid, saddle, and ball-and-socket joints – according to the shapes of the articular surfaces of component bones, and the range of movement they allow. The degree of movement of a particular joint is also affected by how tightly ligaments hold the bones together.

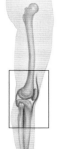

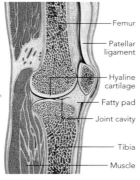

▼ *Movable joint*
*This section though the knee joint shows the main parts of a movable joint. The fatty pad, found in knee and hip joints, has a cushioning effect.*

Femur

Patellar ligament

Hyaline cartilage

Fatty pad

Joint cavity

Tibia

Muscle

▶ **The knee joint** (right) is the largest and most complex joint in the body. As it reaches full extension it rotates slightly and "locks" into a rigid limb from hip to ankle, helped by the bony protuberances at the lower end of the femur and the two semilunar cartilages attached to the sides of the upper end of the tibia. The ligaments surrounding the joint are combined with the cruciate ligaments to prevent overextension. The tendon of the quadriceps femoris muscle surrounds the patella bone in front of the knee. This helps to protect the joint. Behind some tendons and ligaments there are bursae.

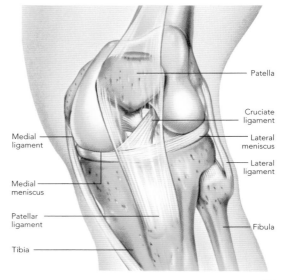

Patella

Cruciate ligament

Lateral meniscus

Lateral ligament

Medial ligament

Medial meniscus

Patellar ligament

Tibia

Fibula

# MUSCLES

Muscles produce movement of, and inside, the body. Muscle tissue is made up of cells called fibers that have the ability to contract, or shorten, in order to produce a pulling force. Muscles are also extensible, and are elastic so that they can be stretched and then recoil and resume their normal resting length. Muscles are also electrically excitable, so that they can be stimulated to contract by a nerve impulse. Three types of muscle – skeletal, smooth, and cardiac – are found in the body.

## Skeletal muscle

As their name suggests, skeletal muscles are attached to the bones of the skeleton. There are more than 640 individually named muscles that together make up more than 40 percent of body weight and give the body its shape. They are arranged in layers that overlap each other. Those just below the skin are called superficial (see below) while those below the superficial muscle are called deep.

Typically each muscle attaches to two or more bones by tough cords of connective

**Skeletal muscle**

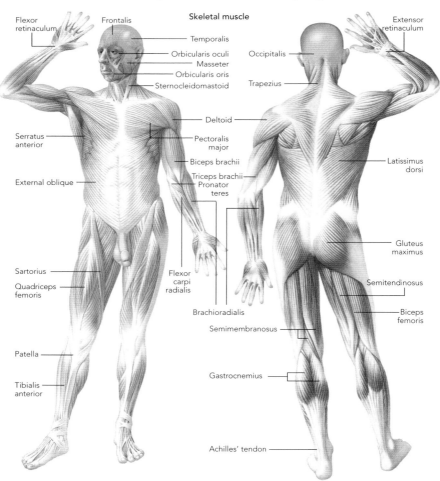

Flexor retinaculum
Frontalis
Temporalis
Orbicularis oculi
Masseter
Orbicularis oris
Sternocleidomastoid
Occipitalis
Trapezius
Extensor retinaculum
Deltoid
Serratus anterior
Pectoralis major
Biceps brachii
Triceps brachii
Pronator teres
Latissimus dorsi
External oblique
Sartorius
Quadriceps femoris
Flexor carpi radialis
Gluteus maximus
Semitendinosus
Brachioradialis
Biceps femoris
Semimembranosus
Patella
Tibialis anterior
Gastrocnemius
Achilles' tendon

tissue known as **tendons**. When skeletal muscles contract across joints, they pull bones, producing a range of movements from running to chewing. Instructions for contraction come from the brain and spinal cord. Skeletal muscle is also known as **voluntary muscle** because a person can make a conscious decision to move their body, although under normal circumstances general body movements occur without conscious involvement. Skeletal muscles also maintain body posture by remaining in a state of partial contraction, or tonus, that holds the body upright; stabilize and strengthen certain joints including the knee and shoulder; and generate heat during contraction that is used to help maintain body temperature at 98.6°F (37°C).

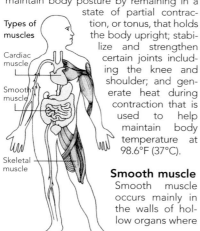

Types of muscles

Cardiac muscle

Smooth muscle

Skeletal muscle

### Smooth muscle
Smooth muscle occurs mainly in the walls of hollow organs where it controls, for example, the movement of food along the small intestine, or the diameter of a blood vessel. It is also known as **involuntary muscle** because its contraction is triggered by nerve impulses from the autonomic nervous system (ANS), and its actions occur without conscious awareness or intervention. In many organs, it forms a double layer with sheets of smooth muscle fibers running at right angles to each other: longitudinal on the outside and circular on the inside. Innervation of the different layers by the opposing parts of the ANS (sympathetic and parasympathetic) causes opposite effects on the organ.

### Cardiac muscle
Cardiac muscle is found solely in the wall of the heart. Its fibers do not require stimulation to contract; they do so automatically, regularly, and without fatiguing, for a lifetime. Electrical impulses from part of the heart, called the pacemaker, travel across and though the heart wall and set the pace of contraction of other cardiac muscle fibers. The sympathetic and parasympathetic branches of the ANS respectively increase or decrease the rate set by the pacemaker in order to meet the human body's changing demands.

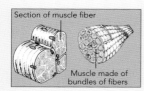

Section of muscle fiber

Muscle made of bundles of fibers

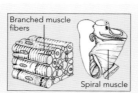

Branched muscle fibers

Spiral muscle

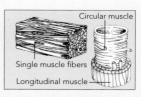

Circular muscle

Single muscle fibers

Longitudinal muscle

▲ *Skeletal muscles are made up of hundreds or thousands of cylindrical fibers, up to 12 inches (30 cm) long, that have a striped (or striated) appearance under the microscope. They are bound together in bundles by connective tissue. They can contract rapidly and powerfully but fatigue (tire) easily.*

▲ *Cardiac muscle fibers, like skeletal muscle fibers, have a striped (or striated) appearance under the microscope, but are shorter and have branches which link up to form interconnected strands. They interweave to form a spiral band around the ventricles of the heart.*

▲ *Smooth muscle fibers are spindle-shaped – wide in the middle and tapering at each end – and are packed together in sheets. They lack the striations seen in other muscle fibers, hence the name 'smooth'. Smooth muscles contract slowly and rhythmically under the control of the autonomic nervous system.*

# SKELETAL MUSCLE – MOVEMENT

Skeletal muscles move the body by pulling the bones that make up the skeleton across the movable joints that connect them.

## Muscle attachment

Each skeletal muscle is anchored to bones at two or more points by **tendons**. Many tendons are cordlike while others, called aponeuroses, are broader and sheetlike. Tendons are fashioned from connective tissue containing closely packed bundles of tough collagen fibers that run in parallel to the direction of pull and provide great tensile strength. The collagen fibers extend through a bone's periosteum (outer membrane) and are firmly embedded in the outer layer of compact bone. Typically, a muscle is attached to two bones: one, the origin, remains fixed during contraction; the other, the insertion, moves toward the origin bone.

## Muscle interactions

Muscles pull but cannot push. When a muscle contracts to produce movement in one direction, there must be another muscle that causes movement in the opposite direction. Skeletal muscles are organized in groups or teams that produce a particular movement or movements. Within teams there are four functional groups of muscles: prime movers, antagonists, synergists, and fixators. A **prime mover**, or agonist, provides the major force for movement, such as the biceps brachii flexing the arm at the elbow. An **antagonist**, such as the triceps brachii which extends the arm at the elbow, opposes the movement produced by the prime mover. **Synergists** assist prime movers by adding some extra force to the movement. **Fixators** are muscles that immobilize a bone or muscle origin, stabilizing that part of the body while other muscles are contracting.

## Muscles, bones, and levers

In terms of basic mechanics, a lever is a rigid bar that moves on a fixed point, called the fulcrum or pivot, when a force, or effort, is applied to it in order to move a weight or load, a principle used in many everyday tools. In biomechanical terms, muscles and bones interact to move the body using lever systems: bones are levers, a joint is a fulcrum, and muscles provide the force to move the load produced by the weight of a body part. Levers fall into three classes according to the relative position of the force, fulcrum, and weight.

## First-class lever

The fulcrum lies between the force and the weight, as in a seesaw. Posterior neck muscles (force) pull the head (weight) up from the chest pivoted on the joint between occipital condyle and atlas (fulcrum).

## Second-class lever

The weight lies between the force and the fulcrum, as in a wheelbarrow. When a person stands on tiptoe, the calf muscles (force) lift the body (weight) pivoting on the ball of the feet (fulcrum).

## Third-class lever

The weight is applied to the lever between the force and the fulcrum, as in a pair of tweezers. During arm flexion, the biceps brachii (force) lifts the forearm (weight) pivoting on the elbow (fulcrum). This is the most common lever system in the body.

## Range of movements

What type of movement, or movements, a muscle produces depend on its location, the type of joint or joints it crosses, and the other muscles it works with. Specific terms are used to describe the movements produced by muscles and bones, as outlined below.

## Flexion and extension

Flexion involves decreasing the angle between two bones, such as the bending of the arm at the elbow. Extension is the opposite action, increasing the angle between two bones, such as the straightening of the arm at the elbow.

## Abduction and adduction

Abduction is the movement of a bone away from the body's midline, such as

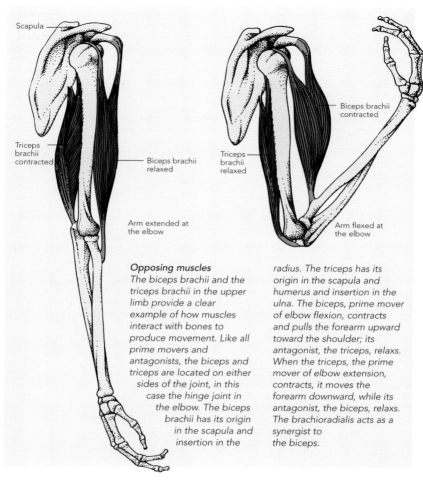

Scapula

Triceps brachii contracted

Biceps brachii relaxed

Arm extended at the elbow

Biceps brachii contracted

Triceps brachii relaxed

Arm flexed at the elbow

*Opposing muscles*
*The biceps brachii and the triceps brachii in the upper limb provide a clear example of how muscles interact with bones to produce movement. Like all prime movers and antagonists, the biceps and triceps are located on either sides of the joint, in this case the hinge joint in the elbow. The biceps brachii has its origin in the scapula and insertion in the radius. The triceps has its origin in the scapula and humerus and insertion in the ulna. The biceps, prime mover of elbow flexion, contracts and pulls the forearm upward toward the shoulder; its antagonist, the triceps, relaxes. When the triceps, the prime mover of elbow extension, contracts, it moves the forearm downward, while its antagonist, the biceps, relaxes. The brachioradialis acts as a synergist to the biceps.*

moving the arm laterally and upward. Adduction is the opposite action, moving a bone toward the body's midline, such as moving the arm medially and downward.

### Elevation and depression
Elevation is an acton that involves lifting upward; for instance, the lifting of the lower jaw in order to close the mouth. Depression is the opposite action; a downward movement, such as opening the mouth.

### Rotation and circumduction
Rotation involves a bone swiveling on its own axis, such as turning the head. Circumduction involves describing a circular movement with the upper or lower limb.

# HOW MUSCLES CONTRACT

Skeletal muscles contract when they receive a nerve impulse from the brain or spinal cord. Contraction is caused by an interaction of protein filaments packed in parallel and in an orderly manner inside muscle fibers (cells). Contraction involves a transformation of chemical energy stored in nutrients, notably glucose, into kinetic (movement) energy, with a concomitant release of heat energy which contributes toward the maintenance of human body temperature at 98.6°F (37°C).

## Skeletal muscle structure

Skeletal muscle consists of hundreds or thousands of long fibers, each between 0.001 mm and 0.1 mm in diameter and up to 12 inches (30 cm) in length, that run in parallel along the length of the muscle. Groups of fibers are grouped together in bundles, called **fascicles**, by an outer layer of connective tissue called **perimysium**. Fascicles, in turn, are bound together by a protective connective tissue sheath, called the **epimysium**, to form a complete muscle. A muscle fiber has a highly organized structure that reflects its unique contractile role. Each fiber is packed with thousands of rodlike **myofibrils** that run in parallel along the fiber. Myofibrils contain two types of protein filaments – thick **myosin** filaments and thin **actin** filaments – that are arranged in repeating units called **sarcomeres**. Within each sarcomere, filaments run in parallel and overlap. This arrangement gives myofibrils and fibers a striated, or striped, appearance when seen under the microscope.

## Muscle contraction

Contraction occurs when the actin and myosin filaments slide over each other. This action requires energy provided in the form of adenosine triphosphate (ATP) released by aerobic respiration inside mitochondria. Inside each sarcomere, myosin filaments are arranged centrally; while actin filaments are attached to the vertical boundary at each end of the sarcomere, called the **Z line**. Extending from each myosin filament are 'heads' that, at rest, project toward but do not touch overlapping actin filaments. These heads are 'charged' with ATP. When a muscle

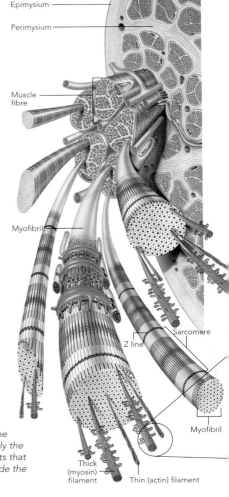

Epimysium

Perimysium

Muscle fibre

Myofibril

Sarcomere

Z line

Myofibril

Thick (myosin) filament

Thin (actin) filament

▶ *Inside a muscle*
*This schematic artwork reveals the hierarchical structure of a skeletal muscle by first sectioning the muscle and then projecting from its cut surface its component muscle fibers, and from them the myofibrils that make up the fibers, and finally the protein filaments that are packed inside the myofibrils.*

fiber is stimulated to contract by a nerve impulse, the myosin head extends toward and touches the actin filament to form a cross bridge. Then, the head uses ATP energy to swivel toward the center of the sarcomere, thereby pulling the acting filament inward. It then breaks away, is recharged with ATP, and makes another cross bridge. This happens repeatedly making the myobril – and therefore the muscle fiber – shorter, until the nerve impulses cease, the cross bridges stop forming, and the muscle fiber relaxes.

## Neurons and muscles
Muscles are stimulated to contract by electrical nerve impulses that travel from the central nervous system along neurons (nerve cells) bundled together in nerves. Where a motor neuron approaches a muscle, it divides into several branches, called **axon terminals**, each of which serves a single muscle-fiber. The motor neuron and muscle fiber meet at a neuromuscular junction, within which the two are separated by a tiny opening, called a **synaptic gap**. Arrival of nerve impulses causes release of a chemical neurotransmitter from the synaptic bulb at the end of the axon ter-

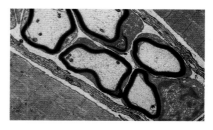

▼ TEM of sectioned nerve fibers (yellow) in skeletal muscle tissue (red). A sheath of myelin (black) surrounds each nerve fiber.

minal. It passes across the gap to the motor end plate, a folded region of the **sarcolemma** (cell membrane of the muscle fiber). Arrival of the neurotransmitter triggers the formation of cross-bridges by myosin and the contraction of the myofibril. Together, each motor neuron and the muscle fibers it supplies is called a **motor unit**. When a neuron fires, it stimulates all the fibers – from tens to thousands of fibers – in its motor unit to contract. The more neurons that fire, and therefore the more motor units that are stimulated, the stronger the contraction of that particular muscle.

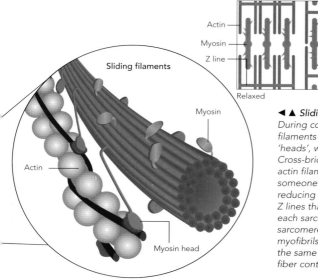

Sliding filaments

Actin
Myosin
Z line
Relaxed
Contracted
Myosin
Actin
Myosin head

◄ ▲ Sliding filaments
During contraction, myosin filaments interact, through their 'heads', with actin filaments. Cross-bridges are formed that pull actin filaments inward (like someone pulling on a rope), reducing the distance between the Z lines that mark the boundary of each sarcomere. With adjacent sarcomeres in neighboring myofibrils becoming shorter at the same time, the entire muscle fiber contracts.

The nervous system is the human body's primary communication system which, together with the endocrine system (see pp. 56–57), controls and coordinates almost all body activities. Working day and night, it receives information from both outside and inside the body, gathered by sensors such as the eyes and proprioceptors; processes and, where appropriate, stores that information; and sends out instructions to effectors, such as muscles and glands, to make the body react. The incredible speed and processing power of the nervous system means that it can cope with a wide range of tasks at the same time. It can, for example, enable a person to think, create, remember, and feel; while, unnoticed, it regulates a host of internal events such as heart rate or body temperature in order to help maintain **homeostasis** – the state of constancy, balance, and stability that normally exists within a healthy body, regardless of external and internal changes.

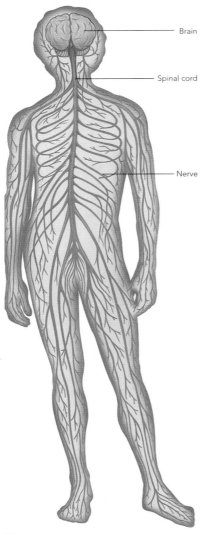

Brain

Spinal cord

Nerve

## Parts of the nervous system

The **central nervous system** (CNS) and **peripheral nervous system** (PNS) are outlined below (see diagram). The PNS has two parts. Its sensory division gathers information from sensors about changes inside and outside the body and transmits it to the CNS. Its motor division carries signals from the CNS to muscles and glands, and itself has two parts: the **somatic nervous system** instructs skeletal muscles to contract, enabling the body to respond consciously; the **autonomic nervous system** or ANS (see pp. 44–45) automatically regulates internal processes, such as breathing and digestion, by transmitting instructions to the body's internal organs.

## Nerve cells

The nervous system is made up of trillions of nerve cells, divided into two main groups: neurons and neuroglia. **Neurons** are long, thin cells that generate and

◄ The nervous system
*The nervous system is divided into two main parts. The central nervous system (CNS) consists of the brain and spinal cord; these form the command center of the nervous system, and carry out the roles of processing and integration. The peripheral nervous system (PNS) consists of the cablelike nerves (bundles of sensory and motor neurons) that extend from the brain and spinal cord and relay messages between all parts of the body and the CNS.*

transmit electrical signals called nerve impulses, and which together form a vast, high-speed communication system. There are three types of neurons. **Sensory neurons** conduct impulses generated when receptors – such as those in the skin, eyes, or muscles – are stimulated to the CNS. **Motor neurons** carry nerve impulses from the CNS to muscles and glands. **Association neurons** or interneurons, which make up 90 percent of all neurons, are found inside the CNS, link sensory and motor neurons, and enable the CNS to sort and process information. **Neuroglia**, also called glial cells, support and nurture neurons.

## Synapses

A synapse is a junction between two neurons. At a synapse is the terminal fiber of an **axon** which reaches out, but does not touch, the dendrite of the next neuron. The small gap between them is called the **synaptic cleft** or synaptic gap. When a neuron is stimulated a nerve impulse – the result of a change in electrical polarity across its membrane – travels at high speed along its axon. The tip of the axon-terminal releases molecules of a chemical called a neurotransmitter, such as acetylcholine. This crosses the synaptic gap in a matter of milliseconds, and depolarizes the membrane of the receiving dendrite, thereby generating a new impulse which passes along the second neuron. This is a one-way process that ensures nerve impulses only travel in one direction. Some association neurons in the brain form hundreds or thousands of synapses with other neurons.

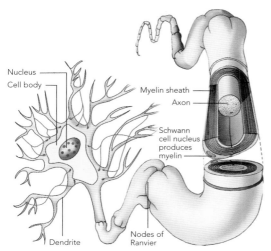

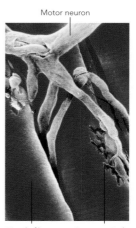

▲ *Neuron*
A neuron divides into three parts: a cell body that contains the cell's nucleus; short, branched processes called dendrites that receive nerve impulses from other neurons and carry them toward the cell body; and the axon, or nerve fiber, typically the longest part of the neuron – it can reach up to 3.3 feet (1 m) in length in parts of the PNS – which carries a nerve impulse to another neuron or a muscle. The axons of some neurons are insulated by a fatty, myelin sheath which increases the speed of transmission of nerve impulses along the neuron, up to 200–250 mph (350–400 kph).

▲ *Synapse* at a neuro-muscular junction. This SEM shows the axon terminal of a motor neuron forming a junction with a muscle fiber. There is a synapse between the terminal and the fiber.

## The forebrain – the cerebrum

Seventy percent of the brain is formed by the two cerebral hemispheres. They are linked together by the nerve fibers running in the **corpus callosum**. The surface area is increased 30 times by the **sulci** and **gyri**. The cortex is densely packed with nerve cells, and the white matter with nerve fibers.

The nerve fibers cross over in the brain stem, so that the right hemisphere is concerned with the left side of the body and vise versa. The left hemisphere is primarily the speech center and mainly controls logical behavior, but has to rely on the right side to assess a three-dimensional world, appreciate artistic values, and recognize friends. The **frontal lobe** has conscious control of movement. The prefrontal area is concerned with intelligence and personality. The **parietal lobe** concerns sensation and body position. The smaller **occipital lobe** assesses vision and the **temporal lobe** assesses hearing. In between are the association areas linking the cerebral cortex.

## The ventricles

Each cerebral hemisphere has a space known as a **lateral ventricle**. The two 'arms' extend forward into the anterior and temporal lobes, and the 'handle' reaches into the occipital region. The lateral ventricles fuse between the two **thalami** to form the third ventricle. A central duct leaves the fourth ventricle to pass down through the spinal cord. The fourth ventricle lies between the cerebellum and brain stem. The ventricular system is filled with cerebrospinal fluid.

## The cerebrospinal fluid

Secreted by the choroid plexus, cerebrospinal fluid is clear and colorless. It bathes the surface of the brain in the subarachnoid space, between the arachnoid and pia mater, and internally in the ventricles, which it joins through a channel in the roof of the fourth ventricle. This channel is known as the **foramen of Magendie**.

It has three functions: to take food to the brain and spinal cord, to remove

The brain

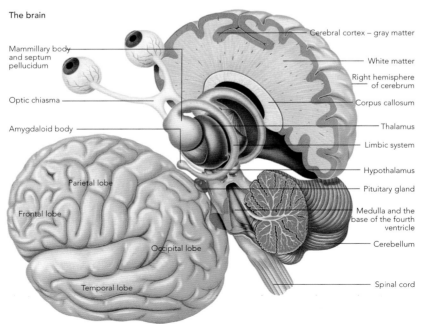

Mammillary body and septum pellucidum

Optic chiasma

Amygdaloid body

Parietal lobe

Frontal lobe

Occipital lobe

Temporal lobe

Cerebral cortex – gray matter

White matter

Right hemisphere of cerebrum

Corpus callosum

Thalamus

Limbic system

Hypothalamus

Pituitary gland

Medulla and the base of the fourth ventricle

Cerebellum

Spinal cord

metabolites, and to act as a shock absorber for the brain.

## The basal ganglia or nuclei

The most important nuclei in the brain are the **caudate** and **lentiform** in the corpus stratum which, with the thalamus and cerebellum, produce smooth, precise movement. They help to coordinate motor movements.

## The limbic system

Consisting of a pair of structures lying above the thalami, the limbic system links the midbrain with the hippocampus and cingulate gyrus and the cerebral cortex. It is concerned with instinctive emotions and memory. The **mammillary body** acts as the initial relay station through the **fornix** to the other areas. The **septum pellucidum** is concerned with pleasurable emotions. The **amygdaloid body** regulates anger and aggression. The **hippocampus** and **cingulate gyrus** help with memory by relating what is happening now with the memory of the past. This helps to maintain the brain's concentration without being distracted by useless information. This inhibits the 'alertness' created by the reticular formation.

## The thalamus

Acting as a relay for sensory information, the thalamus sorts impulses and sends them to the appropriate areas of the parietal lobes. It is also believed to assess sensation.

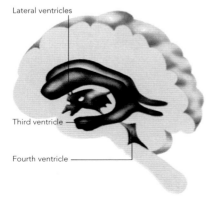

Lateral ventricles

Third ventricle

Fourth ventricle

## The hypothalamus

Lying below the thalamus and above the pituitary gland in the center of the brain, the hypothalamus is partly under the control of the prefrontal lobes and connected with the limbic system. It acts as a link between the nervous and endocrine systems, regulating the autonomic systems as well as the metabolic state of the body.

The hypothalamus is connected with the anterior part of the pituitary gland by a portal venous system beginning and ending in capillaries; this is similar to the hepatic portal system. The blood carries the hormones released in the hypothalamus to control the release of the anterior pituitary hormones. The posterior pituitary gland is controlled by nerves.

The hypothalamus contains centers to control appetite, thirst, body temperature, and libido. There is a fine balance between the amount of food eaten and used. Disturbance of the appetite center may lead to excess appetite and weight gain.

If the blood becomes diluted, the hypothalamus suppresses the production of the antidiuretic hormone (ADH) to allow diuresis to occur. Hemoconcentration will produce the opposite effect with stimulus to produce ADH from the posterior pituitary gland, and the sensation of thirst to stimulate the individual to drink. Libido is due to three factors: instinctive responses, which are genetically inherited; the sex hormones; and sexual behavior and responses learned from infancy. Hunger and thirst will reduce libido through its hypothalamic center.

Temperature regulation is controlled through an assessment of the temperature of the blood. Heat loss can be increased by dilatation of the blood vessels in the skin, convection loss, and sweating; it may be reduced by vasoconstriction and, if necessary, increasing heat production by shivering. The thyroid hormone, thyroxine, has a direct effect on the temperature control.

The activities of the hypothalamus integrate with the rest of the brain's activities but are most closely correlated with those of the brain stem.

# THE BRAIN

### The hindbrain – the cerebellum

The hindbrain is joined directly to the spinal cord. It contains the **medulla** – concerned with sleep, consciousness, breathing, and blood circulation – and the cerebellum.

The **cerebellum** grows rapidly and almost reaches adult size by the age of two. It coordinates the movements initiated by the rest of the brain. This is done by information relayed to it from the medulla oblongata via the nerves in the pons. An assessment is made of the conscious desire; information on balance is received from the ears, visual assessment from the eyes, and other sensory information from the rest of the body. A smooth, regulated movement can then be made, mainly by the inhibition of overreactions.

### The midbrain – the brain stem

The brain stem (*right*) is the vital control area of the brain and is concerned with maintaining all the essential regulatory mechanisms of the body: respiration, blood pressure, pulse rate, alertness, and sleep.

In the **medulla oblongata** – the lower end of the brain stem – the nerve fibers (both sensory and motor) from the spinal cord cross over.

At the upper end there are links with the higher, cerebral centers of consciousness – the limbic and thalamic areas. The nuclei that reflexly control pupil size and eye movements lie in this area of the midbrain.

The third and fourth ventricles with the interconnecting central canal bathe the upper surface with cerebrospinal fluid formed by the choroid plexuses. The foramen of Magendie joins the central canal to the subarachnoid space.

The pons connects the brain stem with the balancing control of the cerebellum. Short nerve fibers – the reticular formation – link the vital centers with each other and monitor the information reaching the brain stem from the sensory tracts of the spinal cord. A change in body position will cause an alteration in blood pressure and pulse rate by control of the diameter of the arterioles. The activity of the reticular formation maintains wakefulness and alertness. If this activity is slowed, for example after a large meal in peaceful surroundings, sleep may occur. Anxiety or fear will increase alertness and raise the blood pressure and pulse rate.

The reticular formation not only controls the vital centers and the production of hydrochloric acid in the stomach, but also helps correlate information from the eyes and ears, and smell and taste from the tongue. The sight and smell of a good meal will cause salivation, increased gastric secretion, and peristalsis. The reticular formation then controls the mechanism of swallowing.

The vomiting center may be activated by nauseating tastes of food or, in some people, by sickening sights. It is easily stimulated by movement – travel sickness – due to the constant changes in the organs of balance.

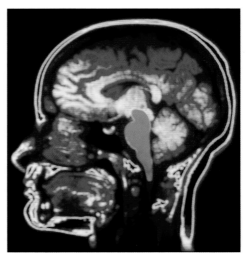

▶ *MRI scan of a human brain seen in side view, with the brainstem area highlighted (blue). The brainstem connects the spinal cord with the upper cerebral regions of the brain (seen folded at top).*

### ▶ Memory

Memory begins with sensation from the body passing to the cerebral cortex. As the information is relayed through the thalamus, it is also passed to the mammillary body and into the limbic system. The stimulus passes through the fornix to the hippocampus and then outward into the diffuse area of the cingulate gyrus. If the memory of a similar stimulus is aroused, the cerebral cortex may be activated.

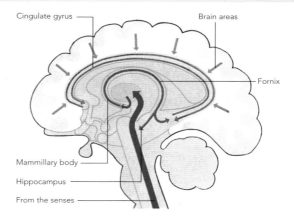

### ▶ Movement

Movement is initiated in the motor cortex and immediately modified by the adjacent suppressor cortical area before direct transmission to the muscles. Each muscular movement is assessed and modified by the cerebellum in conjunction with the thalamus. An unconscious coordination of movement with position, balance, and vision has to be made so that other muscles automatically adjust to give conscious movement.

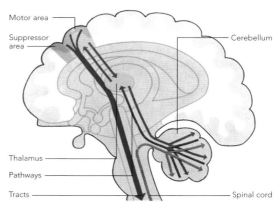

### ▶ Emotion

The emotions are a complex integration of conscious reaction, memory, and instinctive desires. The frontal lobes and limbic systems both affect the hypothalamus, with its centers for anger, thirst, appetite, and sexual desire. These may be stimulated to interact with the motor activity in the brain stem causing alterations in heart, respiratory rate, and muscle tone. The hypothalamus also stimulates the pituitary gland.

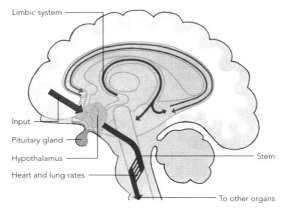

# THE SPINAL CORD

## Spinal cord

The spinal cord provides a vital communication link between the brain and the body by way of paired spinal nerves (spinal nerves are described in more detail on pp. 42–43) that arise from it, relaying information between body and brain; it also processes some incoming information, being involved in many reflex actions (see below). Linked to the brain by the brain stem, the spinal cord is a column of nervous tissue packed with millions of neurons. It has an average length of 18 inches (45 cm) in adults, is no wider that a finger at its midpoint, and extends down the back from the foramen magnum (the large opening in the base of the skull) to the level of the first lumbar vertebra. Just as the brain is protected by the skull, the delicate tissue within the spinal cord is protected by the vertebral foramina of consecutive vertebra which form a bony tunnel, called the **vertebral canal**, along which the spinal cord runs.

## Internal structure

Approximately oval in section, the spinal cord can be divided internally into two clear zones. In its center is a butterfly-shaped mass of **gray matter**. Gray matter contains the cell bodies of motor neurons, which travel from the spinal cord to carry nerve impulses to muscles and glands, the axon terminals of sensory neurons, which carry nerve impulses for receptors, and association neurons, which relay nerve impulses between sensory and motor neurons. Surrounding the gray matter is a cylinder of **white matter**. This consists of nerve fibers, arranged in bundles called tracts, that relay messages between the brain and the peripheral nervous system. Ascending tracts (sensory) carry information about body sensation, which has arrived through sensory neurons in spinal nerves, to the brain; descending tracts (motor) carry instructions from the brain that then travel along the motor neurons in spinal nerves to the body's muscles and glands.

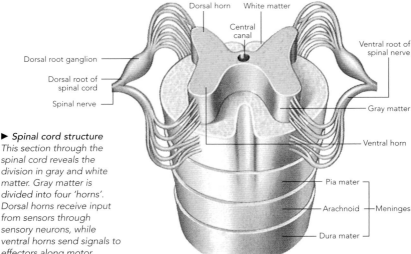

Dorsal horn    White matter

Central canal

Ventral root of spinal nerve

Dorsal root ganglion

Dorsal root of spinal cord

Spinal nerve

Gray matter

Ventral horn

Pia mater
Arachnoid — Meninges
Dura mater

▶ *Spinal cord structure*
*This section through the spinal cord reveals the division in gray and white matter. Gray matter is divided into four 'horns'. Dorsal horns receive input from sensors through sensory neurons, while ventral horns send signals to effectors along motor neurons. The diagram also shows the three meninges, the membranes that surround and protect the spinal cord, and the two roots of each spinal nerve* *that emerge from the spinal cord. The dorsal root contains sensory neurons which carry incoming nerve impulses – the cell bodies of* *sensory neurons are located in the dorsal root ganglion – while the ventral root contains motor neurons which carry outgoing signals.*

## Meninges

Three cylinders of connective tissue, called the meninges, surround the spinal cord. On the outside is the **dura mater**, below which is the **arachnoid**, which, in turn, surrounds the **pia mater**. Circulating in the subarachnoid space – between the arachnoid and the pia mater – and in the spinal cord's central canal, is a colorless, watery liquid called **cerebrospinal fluid** (CSF). CSF has two roles: it acts as a shock absorber, cushioning the spinal cord from knocks and shocks; and it delivers nutrients from, and removes wastes to, the bloodstream.

## Reflexes

Reflexes are automatic, involuntary, unchanging, split-second responses, many of which protect the body from hazards. A simple example is the withdrawal reflex by which we automatically pull away our hand when it touches a hot or sharp object. Other reflexes include

*Reflex action*
*If a person touches something hot, sensory receptors in the skin generate nerve impulses that travel to the spinal cord along sensory neurons. In the spinal cord, messages are relayed by association neurons to motor neurons that transmit nerve impulses to the upper arm muscles, which withdraw the hand from the source of the heat.*

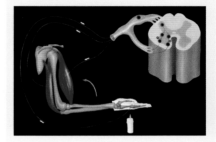

blinking and swallowing. Many reflexes involve the spinal cord, since this proves the shortest, and fastest, pathway to a response. This short pathway is called a **reflex arc**. It typically involves a receptor responding to a stimulus by generating an impulse that travels along sensory neurons, association neurons, and motor neurons to a skeletal muscle. Only after the rapid, reflex action happens do nerve impulses arrive in the brain, making the person consciously aware of the stimulus, perhaps feeling pain or crying out.

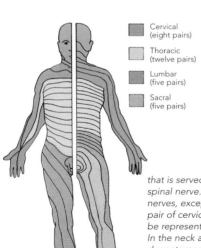

Cervical (eight pairs)

Thoracic (twelve pairs)

Lumbar (five pairs)

Sacral (five pairs)

▲ **Dermatomes**
*A dermatome ('skin segment') is an area of skin* *that is served by a single spinal nerve. All spinal nerves, except for the first pair of cervical nerves, can be represented in this way. In the neck and trunk, dermatomes form horizontal bands that are in line with their respective nerves. In the upper and lower limbs, the dermatomes are aligned lengthwise. Doctors can use a dermatome map to identify damaged nerves. A* *lack of sensation in one particular skin area may enable the doctor to trace the problem back and identify the specific spinal nerve, possibly injured, that should normally serve this area. This dermatome map of the body divides up the skin's surface in terms of its innervation. Color-coding indicates whether a dermatome is served by cervical, thoracic, lumbar, or sacral, spinal nerves.*

# NERVES

These creamy white, glistening cords provide the information highways that arise from the central nervous system, or CNS (brain and spinal cord) and relay nerve impulses between the CNS and all regions of the body, including sense organs, muscles, and glands. Each nerve consists of parallel bundles of hundreds or thousands of neurons or, more specifically, the long nerve fibers of those neurons. The vast majority of nerves, called **mixed nerves**, carry a two-way traffic of nerve impulses because they contain the fibers of both motor and sensory neurons – purely sensory nerves or motor nerves are rare.

Inside a nerve, each bundle of neurons, or **fascicle**, is surrounded by a connective tissue sheath, the **perineurium**. Another connective tissue sheath, the **epineurium**, holds the fascicles together, along with their blood vessels. The epineurium protects the axons but permits bending of the nerve when the body moves. The largest, and longest, nerve, the **sciatic**, which supplies the leg, is about 1 inch (2 cm) wide where it emerges from the lower part of the spinal cord. The thinnest nerves are no wider than a hair. Nerves are divided into two groups – spinal nerves and cranial nerves – according to which part of the CNS they arise from.

## Spinal nerves

Thirty-one pairs of spinal nerves fan out in a repeating pattern from the spinal cord through openings between adjacent vertebrae called intervertebral foramina. All are mixed nerves. Their sensory neurons collect general sensory information from, for example, receptors in the skin relating to touch and pain, while their motor neurons carry instructions that control most of the body's skeletal muscles, as well as smooth muscles and glands. Spinal nerves divide into four groups (see illustration, p. 41): eight pairs of **cervical nerves** supply the neck, upper limbs, and the diaphragm; twelve pairs of **thoracic nerves** supply muscles in the

**Spinal nerves**

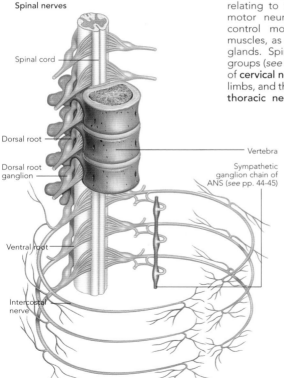

Spinal cord

Dorsal root

Dorsal root ganglion

Ventral root

Intercostal nerve

Vertebra

Sympathetic ganglion chain of ANS (see pp. 44-45)

◀ *Spinal nerves This is a diagram of the thoracic region of the backbone and spinal cord. It shows how spinal nerves are formed by the merging of a dorsal root (carrying sensory nerve fibers) and a ventral root (carrying motor nerve fibers), both arising from the spinal cord. The dorsal root ganglion (pl. ganglia) is a swelling that contains the cell bodies of sensory neurons. The twelve intercostal nerves (the thoracic spinal nerves) supply the skin and muscles of the thorax and much of the abdominal wall.*

chest and back; five pairs of **lumbar nerves** supply the lower abdomen and parts of the lower limbs; five pairs of **sacral nerves** (and one pair of **coccygeal nerves**) supply reproductive organs and the bladder, and other parts of the lower limbs. In the upper and lower parts of the spinal cord, spinal nerves merge to form an interconnecting nerve network called

**Nerve structure**

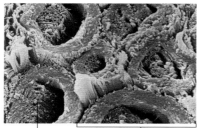

Bundle of axons
(nerve fibers)

Fascicle bound by
perineurium

▲ *Colored SEM of a section through bundles of nerve fibers (green). Each fiber consists of a nerve cell axon, the output process of a nerve cell, surrounded by a fatty insulating layer known as a myelin sheath. Myelin sheaths increase the transmission speed of the electrical nerve signals.*

▶ *Cranial nerves There are twelve nerves, I–XII (right), which arise from the under-surface of the brain. They supply the head, neck, and major organs in the body. Sensory fibers are shown in blue, motor fibers in red. Three are solely sensory: olfactory I, optic II, and vestibulocochlear VIII. Two are mixed nerves, but are primarily motor in function: accessory XI to neck muscles, and hypoglossal XII to the tongue. Of the remaining mixed nerves: oculomotor III, trochlear IV,*

a **plexus**. The cervical plexus in the neck, and the brachial plexus in the neck and axilla, control the upper limbs; the lumbar plexus in the lower back, and the sacral plexus in the pelvic region, control the lower abdomen and lower limbs.

## Cranial nerves

Twelve pairs of cranial nerves arise from the brain and emerge through foramina in the skull. The cranial nerves serve only the head and neck, apart from the **vagus nerves**, which extend into the thorax and abdomen. Most cranial nerves are mixed, carrying nerve impulses from sensors to the brain, or carrying instructions from the brain to the muscles of the head and neck; three pairs, linked to special sense organs, are purely sensory. As well as being named, each pair of cranial nerves is also numbered using Roman numerals.

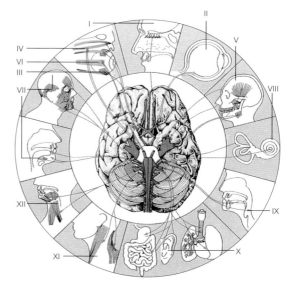

*and abducens VI supply the eye muscles; trigeminal V involves chewing and facial sensations; VII facial involves facial expression and taste in the anterior*

*two-thirds of the tongue; glossopharyngeal IX involves swallowing and the remainder of taste; vagus X supplies chest and abdominal organs.*

# AUTONOMIC NERVOUS SYSTEM

The autonomic nervous system (ANS) is the section of the nervous system that, second by second, automatically regulates involuntary functions – such as heart rate, digestion, pupil size – in order to help in the process of homeostasis (maintaining a stable environment inside the body) and to enable the body to respond to the changing external and psychological demands on the body by causing arousal or quiescence.

## How the ANS works

The ANS is a purely motor system with axons (nerve fibers) of motor neurons running from the central nervous system (CNS) to target structures: smooth (invol-untary) muscles in the walls of body organs, such as the bladder, small intestine, and blood vessels; to glands such as the adrenal and salivary glands; and to the cardiac muscle in the heart. Changes in the internal and external environment are detected by sensors that send a constant stream of impulses along sensory neurons to the CNS, notably to centers in the brain stem which are, in turn, regulated by the hypothalamus, the ultimate controller of the ANS. Incoming information is processed in the CNS, and signals are sent out along the motor neurons of the ANS to target organs.

## Two divisions

The ANS has two divisions – sympathetic and parasympathetic – that have opposite effects. Target organs receive a supply of nerve fibers from both of them. The two antagonistic ANS divisions provide a self-regulating, dynamic balance that can be shifted, when necessary, to meet the body's demands. The **sympathetic division** is generally excitatory, prepares the body for stress, and encourages energy expen-diture. Sympathetic dominance increases heart rate and blood pressure, increases breathing rate, increases blood flow to the muscles, widens the pupils, and stimulates

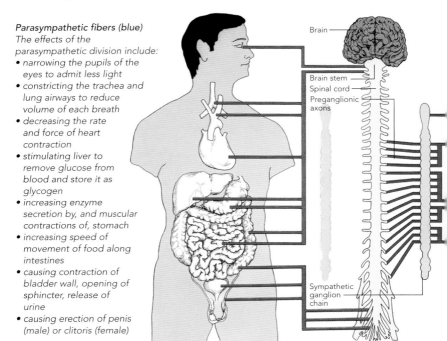

*Parasympathetic fibers (blue)*
*The effects of the parasympathetic division include:*
- *narrowing the pupils of the eyes to admit less light*
- *constricting the trachea and lung airways to reduce volume of each breath*
- *decreasing the rate and force of heart contraction*
- *stimulating liver to remove glucose from blood and store it as glycogen*
- *increasing enzyme secretion by, and muscular contractions of, stomach*
- *increasing speed of movement of food along intestines*
- *causing contraction of bladder wall, opening of sphincter, release of urine*
- *causing erection of penis (male) or clitoris (female)*

Brain

Brain stem
Spinal cord
Preganglionic axons

Sympathetic ganglion chain

the liver to release energy-rich glucose into the blood. The sympathetic division also stimulates the adrenal glands to release epinephrine and norepinephrine, two hormones that mimic and prolong the effects of the sympathetic division. The **parasympathetic division** restores and conserves resources when the body is resting or recovering. Parasympathetic dominance produces calm and quiescence by slowing heart and breathing rates, and by prioritizing digestion in order to take in supplies of raw materials and energy, and to store glucose in the liver ready for the next stressful episode.

Nerve fibers (axons) of the sympathetic division leave the thoracic and lumbar sections of the spinal cord through spinal nerves; nerve fibers of the parasympathetic division leave the brain stem through cranial nerves and the spinal cord through spinal nerves. Unlike the somatic nervous system – where single motor neurons run from the CNS to the

target skeletal muscle – in the ANS two motor neurons runs in series, one following the other. The axon of the first neuron (called the **preganglionic axon**) runs from the CNS to a swelling called a **ganglion**, which contains a mass of axon terminals and acts as a relay station. The axon of the second neuron (**postganglionic axon**) runs from the ganglion to the target muscle or gland. In the sympathetic division, ganglia are linked in a vertical sympathetic chain, close to and on either side of the spinal cord. In the parasympathetic division, ganglia lie close to or within the target organ. The two divisions have opposite effects on target organs because their nerve endings release two different neurotransmitters which have an excitatory or inhibitory effect depending on the organ: sympathetic nerve endings release norepinephrine, and parasympathetic nerve endings release acetylcholine.

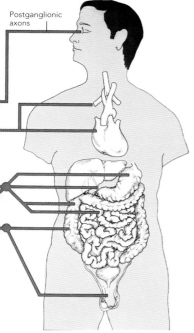

Postganglionic axons

*Sympathetic fibers (red)*
*Effects of the sympathetic division include:*
- *widening the pupils to admit more light*
- *widening the trachea and lung airways to increase the volume of each breath*
- *increasing the rate and force of heart contraction*
- *decreasing enzyme secretion by, and muscular contractions of, stomach*
- *stimulating liver to release glucose into blood*
- *decreasing speed of movement of food along the intestines*
- *causing relaxation of bladder wall, closing of sphincter, thereby inhibiting release of urine*
- *causing ejaculation (male) and contraction of vagina (female)*

◄ *Autonomic Nervous System*
*The two divisions of the ANS spread out from both sides of the brain stem and/or spinal cord. For purposes of clarity, only half of each division is shown in this diagram. Some of the effects of the two divisions are described above.*

# TOUCH AND PAIN

Without a constantly updated input of information about conditions inside or outside the body, it would be impossible for the brain to function. This information is provided by a range of sensory receptors. These detect stimuli – changes in their local environment – and generate nerve impulses which are transmitted along sensory neurons to the central nervous system (CNS), and then to the relevant sensory area of the brain. Many receptors, such as those for taste, smell, sight, hearing, and balance – the five 'special senses' (described in detail on pp. 48–55) – are located in discrete sense organs (tongue, nose, eyes, and ears) in the head. Touch, on the other hand, is known as a general sense because its receptors are scattered throughout the body.

## Touch

Touch is a blanket term used to describe the sensations evoked by a range of receptors in the skin. The skin, as already described (pp. 12–13), provides a large sensory interface between the body and its surroundings that enables humans to perceive a broad range of sensations including the chill of a dip in an icy sea, the pain of a paper cut, the softness of velvet, or the pressure caused by holding a pen between finger and thumb. Perception, as with any sense, occurs when a person becomes aware of the sensation, following processing by the cerebral cortex of the 'raw' input of nerve impulses from the receptor.

Sensory receptors involved in touch differ according to the stimuli to which they respond. **Mechano-recep-** tors generate nerve impulses when they are pulled, pushed, or squeezed. They are also found as **proprioceptors**, or stretch receptors, in muscles and tendons where they detect tension and send messages to the brain that enable it to maintain the body's posture. **Thermoreceptors** respond to cold or heat. **Nociceptors** respond to painful stimuli.

## Receptors in the skin

The majority of sensory receptors in the skin are mechanoreceptors – their roles are related to their relative 'depth' in the dermis. **Merkel's disks**, which extend into the epidermis near to the skin's surface, detect light touch and pressure, helping to sense its source. **Meissner's corpuscles**, located just below the epidermal/dermal interface, also sense light touch. Deeper in the dermis, **Ruffini's corpuscles** respond to continuous touch on the skin. Deeper still, the larger **Pacinian corpuscles** detect vibration and firm, heavy pressure. All of these mechanoreceptors are encapsulated, that is enclosed, by a membrane. The remaining receptors are free nerve endings that detect movement of hairs (also mechanoreceptors), heat and cold, and also pain.

Nerve impulses from skin receptors are carried by sensory neurons – via the spinal nerves, spinal cord, brain stem, and thalamus – to the sensory area of the

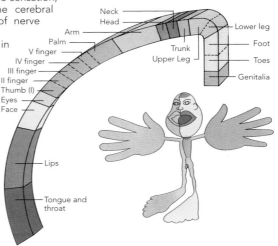

Neck
Head
Arm
Palm
V finger
IV finger
III finger
II finger
Thumb (I)
Eyes
Face
Lips
Tongue and throat

Lower leg
Foot
Trunk
Upper Leg
Toes
Genitalia

▶ *Homunculus This odd representation of the body, called a homunculus, shows the relative amount of brain power in the sensory cortex that is devoted to the variable numbers of sensory receptors in different parts of the body.*

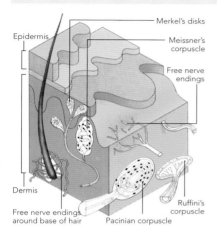

Epidermis
Merkel's disks
Meissner's corpuscle
Free nerve endings
Dermis
Ruffini's corpuscle
Free nerve endings around base of hair
Pacinian corpuscle

▲ *Skin receptors This section through the skin shows its two layers, the thinner protective epidermis and the thicker dermis. Most sensory receptors are located in the dermis, although some may extend into the epidermis.*

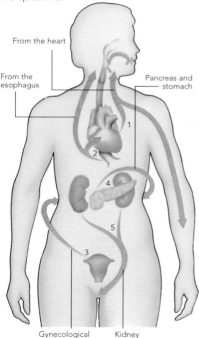

From the heart

From the esophagus

Pancreas and stomach

1

2

4

5

3

Gynecological        Kidney

cerebral cortex. Touch receptors are not evenly distributed over the skin. Some areas – such as the fingertips and lips – have many more receptors than others – such as the back – making them far more sensitive. This explains why blind people can use their fingers to 'read' Braille script. Because of this uneven distribution, a larger portion of the sensory cortex is needed to deal with sensory input from, say, the fingers, than from the elbow.

## Pain

Pain receptors, or **nociceptors**, are found not only in the skin but in most other parts of the body – including joints, the stomach, and muscles – apart from the brain. They are stimulated by damage to tissues, typically by chemicals released by damaged cells, or by intense pressure or heat. Pain serves a useful purpose by alerting the body to actual or potential tissue damage. Somatic pain arises from skin, muscles and joints, can be either superficial – a sharp pain that can make someone cry out – or deep – an aching or burning feeling – that comes from muscles, joints, or deep in the skin. Visceral pain (*see* referred pain below) is a dull ache or gnawing pain caused by the stretching or irritation of nociceptors in thoracic and abdominal organs.

◄ *Referred pain*
*Internal organs and structures are well supplied with nerves, but pain is diffuse and poorly located compared with skin sensation (below). Most of the pain is produced by stretching and contracting, hence the pain of colic. Internal pain will cause stimulation of local nerves in a segment of the spinal cord; this makes it appear that the pain is coming from the skin, which is supplied by the sensory nerves. The heart (1) and esophagus (2) refer pain to the neck, shoulders and arms; uterus (3) and pancreas (4) to the lumbar region; and kidneys (5) into the groin. Diaphragmatic pain may be referred to the shoulders, as the phrenic nerve of the diaphragm is formed from the spinal nerves in the neck, which also supply the shoulders.*

# TASTE AND SMELL

Taste and smell are linked senses. First, they are located in close proximity to each other. Second, they complement each other and enable human beings to perceive flavors. Third, they both utilize the same type of sensory receptors, called **chemoreceptors**. Chemoreceptors contain microscopic 'hairs', which detect specific chemicals in food or drink, or in the air and generate nerve impulses which are sent to the relevant region of the brain so they can be perceived.

## Taste

The tongue is the organ of taste. This highly mobile, muscular organ moves food while it is being chewed and mixed with saliva, and also aids in the formation of words during speech. The tongue's surface is covered by tiny projections called **papillae**, which give it a characteristic bumpy appearance. There are three types of papillae: spiky **filiform papillae** cover the tongue and give its surface a rough texture to grip food during chewing and swallowing; mushroomlike **fungiform papillae** are found near the edges

and tip of the tongue; between seven and 10 large **circumvallate** (or vallate) **papillae** form a V-shape at the back of the tongue.

Housed in the sides of fungiform and circumvallate papillae are some 10,000 receptors called **taste buds**. Each consists of between 40 and 60 sensory and supporting cells arranged like the segments of an orange. Tiny 'hairs' (taste hairs), called **microvilli**, project from sensory cells into the opening of the taste bud, called the **taste pore**, on the side of the papilla. When food is chewed, its constituent chemicals, such as sugars, dissolve in saliva and flow into the taste pore and make contact with receptors on the taste hairs. These then generate nerve impulses that travel to the gustatory (taste) area of the cerebral cortex where tastes are perceived. Taste buds can detect just four tastes – sweet, sour, salt, bitter – but when combined with sensory data from the nose, the brain can distinguish between a wide range of flavors. The tongue also contains receptors for touch and pressure – allowing the texture

▼ *Tongue and taste buds*
*This shows a detailed view of the surface of, and a section through, the tongue from a posterolateral view.*

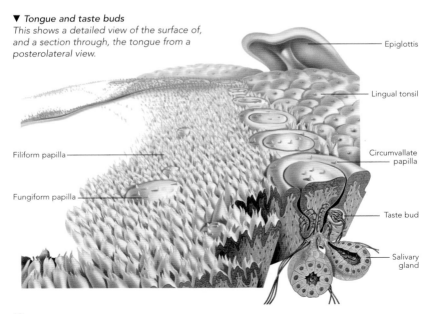

Epiglottis

Lingual tonsil

Filiform papilla

Fungiform papilla

Circumvallate papilla

Taste bud

Salivary gland

of food to be determined – and for heat and cold – allowing the eater to distinguish between hot potatoes or ice cream. Pain receptors detect chemicals in spicy 'hot' food such as chilies.

## Smell

Smell receptors are located in a postage-stamp sized area (around 0.4 inches/ 2.5 sq cm) in the upper parts of the nasal cavity on either side of the basal septum. This area of olfactory epithelium contains around 10 million olfactory receptors, from which project between six and 20 hairlike cilia that detect odor molecules. When a person breathes in, air is carried high into the nasal cavity where the odor molecules it is carrying dissolve in watery mucus. Once dissolved, the odor molecules attach themselves to the cilia, which generate nerve impulses that are carried to the temporal lobe of the cerebral cortex (where they are perceived as smells), as well as to the limbic system, which, since this part of the brain deals with emotions and memories, explains why certain smells can evoke such power-

▼ *Olfactory receptors* This SEM shows cilia (colored orange) projecting from an olfactory receptor in the nasal cavity.

ful memories. The sense of smell is 10,000 times more sensitive than taste, and can detect more than 10,000 different odors. There appear not to be 10,000 different receptor sites on the cilia of olfactory receptors. Instead, it seems that each smell is made up of one or more components from a set of 100 or so basic smells. It is to these basic smells that receptors respond.

Together, taste and smell allow humans to detect a range of flavors, which enhances, for instance, the enjoyment of a good meal. Of the two, smell is the 'senior' partner, as becomes clear when someone has a heavy cold and cannot discern any flavor at all. The perception of smells and/or tastes automatically causes the release of saliva from the salivary glands, and gastric juice from gastric glands, to aid in the process of digestion. Smell and taste also allow us to avoid harm by, for example, early detection of burning or being able to spit out food because it smells and tastes spoiled.

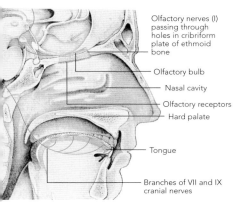

Olfactory nerves (I) passing through holes in cribriform plate of ethmoid bone

Olfactory bulb

Nasal cavity

Olfactory receptors

Hard palate

Tongue

Branches of VII and IX cranial nerves

▲ *Nerve pathways* This section through the head shows the routes taken by sensory fibers from olfactory and taste receptors to the brain. Nerve fibers of olfactory nerves (cranial nerve I) pass from olfactory receptors through tiny holes in the ethmoid bone (which forms the roof of the nasal cavity) and synapse with fibers in the olfactory bulb, an anterior extension of the brain. In the tongue, a branch of the facial nerve (cranial nerve VII) carries impulses from the anterior two-thirds of the tongue to the brain. A branch of the glossopharyngeal nerve (cranial nerve IX) carries impulses from the posterior third of the tongue to the brain.

# EYES AND VISION

Vision is the body's most important sense. The organ that detects light, the eye, is one of the most complex in the body, and the 125 million photoreceptors it contains – receptor cells that generate nerve impulses when stimulated by light – make up 70 percent of the sensory receptors in the body. Light entering the eye is focused on photoreceptors at the back of the eyeball, rather like a camera focusing light onto a film. Nerve impulses generated in the eyes are carried by optic nerves to the brain where they are processed in order that we can perceive visual images.

## The eyeball

Each eyeball is about 1 inch (2.5 cm) in diameter. It is cushioned by pads of fat, and is located within, and protected by, a bony socket, or orbit, in the skull such that only the anterior one-sixth of the eye is visible from the outside. The exposed front of the eye is protected from dust and excessive light by eyelashes, and by the eyelids that open and close regularly during blinking and sweep tears across the eye's surface. Produced by **lacrimal** **glands** located just above the eye, and released through tear ducts onto the eyeball's surface, tears wet the front of the eye, wash away dust particles, and kill bacteria, before draining through two openings in the inner corner of the eye into the nasal cavity.

## Eye structure

The simplest way to examine its structure is to look at the three layers from which the eyeball is made (*see* diagram below). The outer layer consists of the tough, slippery **sclera**, which covers most of the eyeball and forms the visible white of the eye; and the **cornea**, the clear domed region that allows light to enter the eye. The middle vascular layer, or **uvea**, consists, in much of the eyeball, of the **choroid**. This dark layer supplies the other layers with blood, and because of its blackness prevents the internal reflection of light. Anteriorly it thickens to form the **ciliary body**, which contains a ring of muscle. Projecting inward from the ciliary body are the suspensory ligaments that hold the transparent **lens**, made from fibers containing the proteins called crys-

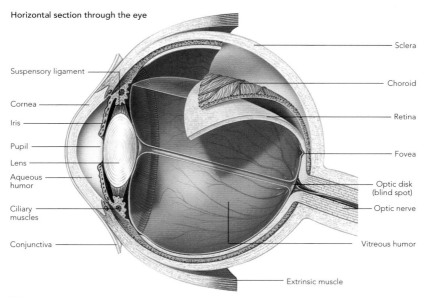

Horizontal section through the eye

Suspensory ligament

Cornea

Iris

Pupil

Lens

Aqueous humor

Ciliary muscles

Conjunctiva

Sclera

Choroid

Retina

Fovea

Optic disk (blind spot)

Optic nerve

Vitreous humor

Extrinsic muscle

tallins, in place. The **ciliary muscles** adjust the shape of the lens to fine focus light already partly focused by the cornea. The vascular layer projects inward, anterior to the ciliary body, to form the **iris**. The iris contains melanin pigment which, depending on its concentration, gives the eye a color ranging from pale blue to dark brown. The iris also contains smooth muscle fibers, arranged in radial and circular fashion. The fibers automatically regulate the size of the central opening or pupil to control the amount of light entering the eye (*see below*).

The inner layer of the eye, the **retina**, covers the choroid and extends anteriorly to cover the ciliary body. It consists of an array of photoreceptors called rods and cones (*see also* pp. 52–53), and the nerve fibers that carry nerve impulses generated by the photoreceptors to the brain. The 120 million **rods** work best in dim light and produce black-and-white images. The five million **cones** work best in bright light, detect colors, and provide the greatest detail. Rods are dispersed throughout the retina, while cones are concentrated in the **fovea**, the central

point of the retina directly behind the lens. Retinal nerve fibers leave the eyeball through the optic nerve. At its exit point, the **optic disk**, there are no photoreceptors. This 'blind spot', where no light is detected, leaves a visual gap that is 'filled' by the brain.

Either side of the lens, the eye divides into two unequal cavities. The larger and posterior cavity is filled with the gelatin-like vitreous humor. The smaller and anterior cavity is filled with watery aqueous humor. Both under slight pressure, the humors serve to preserve the shape of the eyeball.

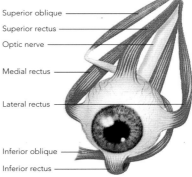

Superior oblique
Superior rectus
Optic nerve
Medial rectus
Lateral rectus
Inferior oblique
Inferior rectus

▲ *Eye muscles*
*The eyeball is moved by six extrinsic, or external, straplike muscles. The movements of both eyes are coordinated by the brainstem so that they do not move independently in different directions. The superior rectus makes the eye look upward. The inferior rectus makes the eye look downward and inward. The medial rectus makes the eye look medially. The lateral rectus makes the eye look laterally. The superior oblique swivels the eye so it looks downward and outward. The inferior oblique makes the eye look upward and outward.*

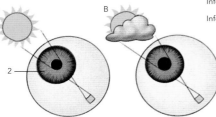

▲ *Pupil reflex*
*The retina is very sensitive to light. Too much light (**A**) distorts color and is dazzling. The pupils vary in size and thus reduce or increase the amount of light entering the eye. Bright light causes a reflex nervous reaction, controlled by centers in the midbrain. The circular pupillary muscle (**1**) in both irises contracts and the radial fibers (**2**) extend, thus narrowing the diameter. Poor light (**B**) will make both pupils dilate, allowing sufficient light to stimulate the cells in the retina (**3**).*

# EYES AND VISION

## How the eye works

When light enters the eye from an object, it is refracted (bent) first by the cornea and then by the lens to produce an inverted, clearly focused image on the retina that will enable the brain to produce a clear view of the outside world. Fine adjustment to focusing is the role of the **lens** because, unlike the cornea, it has elasticity and can change shape in order to focus light from any object whether it is nearby or far in the distance. When looking at distant objects, the ring of **ciliary muscle** around the lens relaxes. The **vitreous humor** pushes outward to increase the diameter of the ciliary body, causing the suspensory ligaments to pull on the lens and make it flatter, thereby providing the right shape to focus the nearly parallel light rays from the faraway object. When looking at an object nearby, the ciliary muscle contracts thereby reducing its diameter, the suspensory ligaments become less taut, the lens reverts to its natural, rounder shape in order to focus the divergent light rays arriving from the nearby object. Changing lens shape, known as **accommodation**, is controlled automatically by the brain. Short-sightedness, or **myopia**, is caused by light rays being focused in front of the retina; long-sightedness or **hypermetropia**, is caused by light rays being focused behind the retina. Both can be corrected by glasses or contact lenses.

Focused by cornea and lens, light falls on the photoreceptors in the retina. These contain a visual pigment that undergoes a change when hit by light, and causes the generation of a nerve

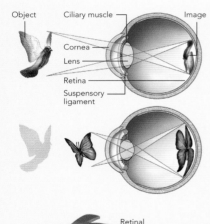

Object | Ciliary muscle | Image
Cornea
Lens
Retina
Suspensory ligament

◀ *Far and near*
*These are the two images showing a section through the eyeball to show how the lens changes shape to focus light from far and near objects.*

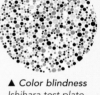

▲ *Color blindness*
*Ishihara test plate. Color blind people will not be able to see the spade and fork.*

▼ *Rods and cones This detailed view of the retina shows the arrangement of rods (blue) and cones (green), and their links to the nerve fibers which leave the eye through the optic nerve. It begins with the section through the eyeball showing light being focused on the fovea, with a zoom to a section through the retina, and the detail of a rod and a cone.*

Retinal nerves
Retinal nerve fibers | Connecting neurons
Direction of light
Direction of nerve impulses
Cone
Rod

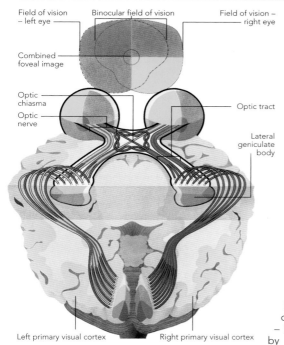

Field of vision – left eye

Binocular field of vision

Field of vision – right eye

Combined foveal image

Optic chiasma

Optic nerve

Optic tract

Lateral geniculate body

Left primary visual cortex

Right primary visual cortex

**◄ Visual pathways**
*Section though the eyes and brain to show the route taken by nerve impulses from the retina to the visual cortex.*

cortex, and nerve impulses from the left half of each eyeball travel to the left primary visual area. They do so, on each side of the brain, along an optic tract which passes to the **lateral geniculate body**. This body relays nerve impulses to the primary visual areas, where movement and outlines are identified. Other cerebral areas refine the image so that it can be perceived. The slightly different inputs received from the left eye and the right eye enables the brain to construct a three-dimensional view of the world – its binocular field of vision – by combining the two inputs. This binocular, or stereoscopic, vision enables us to judge depth and distance accurately.

impulse which is transmitted to the visual cortex in the posterior region of the cerebral hemispheres. As already described (see pp. 50–51) there are two types of photoreceptors: rods and cones. **Cones** detect colors and provide the greatest detail and are concentrated mainly in the **fovea**, the point on which light falls when an object is looked at directly.

## Visual pathways

Nerve impulses are carried away from the retina by nerve fibers that synapse with the photoreceptors. Because the eyes face forward, they see the same view from slightly different angles, so the brain's visual processing area – the primary visual cortex – also receives slightly different sets of visual input. Nerve fibers travel along the optic nerves to a point of crossover called the **optic chiasma**. Here the nerve fibers partly cross over so that nerve impulses from the right half of each eyeball travel to the right primary visual

## Color blindness

Color vision depends on the presence of three types of cones that each respond to a particular color or wavelength of light: red, green, or blue. About 10 percent of men and one percent of women (this is a sex-linked characteristic) have an inherited disorder called color deficiency, or color blindness, caused by the absence of one or more types of cone. This is commonly a red or green deficiency, resulting in an inability to distinguish between those two colors. A blue deficiency is rarer, as is total color blindness. Defects in color vision can be detected using Ishihara test plates (see picture opposite). The spade is purple and thus stimulates blue and green cones, and cannot be seen by the green color blind. The fork is red so it is not seen by the red color blind. Both are seen by those with normal color vision.

# HEARING

The ear is the organ of hearing, and also plays a key role in the body's sense of balance. What most people refer to as the ear is, in fact, the outer flap, or pinna, that projects from the side of the head. The main part of the ear is concealed and protected within the temporal bone of the skull. The ear is divided into three distinct sections: the outer ear and middle ear, both filled with air; and the fluid-filled inner ear.

## Outer ear

The outer ear consists of the **pinna**, which is reinforced with elastic cartilage, and the 1 inch (2.5 cm) -long **auditory canal**. The outer ear serves to gather sound waves and direct them into the ear. Ceruminous glands in the skin, lining the auditory canal, produce a thick secretion called cerumen, or earwax, which, together with small hairs, keeps the canal clean and repels insects. Separating the outer and middle sections of the ear is a tightly stretched piece of skin called the **eardrum** or tympanic membrane.

## Middle ear

Extending across the middle ear, from the **cochlea** to the oval window, are three tiny bones, or **ossicles**, named in Latin or English for their shapes. The **malleus** (hammer) is attached to the eardrum; the **stapes** (stirrup) is attached to the oval window, the membrane-covered opening to the inner ear; the central **incus** (anvil) connects the malleus and stapes through free-moving synovial joints.

## Inner ear

Also called the labyrinth, the inner ear has two major parts. The **bony labyrinth** consists of channels and cavities that twist and turn through the temporal bone. These channels and cavities are filled with a liquid called **perilymph** and have three functional regions: the **cochlea**, which is involved in hearing; and the **vestibule and semicircular canals** which are involved in balance (see below). The **membranous labyrinth** is a continuous series of membranous ducts that follows the contours of the bony labyrinth and floats in **perilymph**. The ducts themselves are filled with a fluid called **endolymph**.

The snail-shaped cochlea is divided by two membranes – the vestibular membrane and the basilar membrane – into three parallel channels: the **vestibular duct** runs from the oval window; the **tympanic duct** ends at the round window,

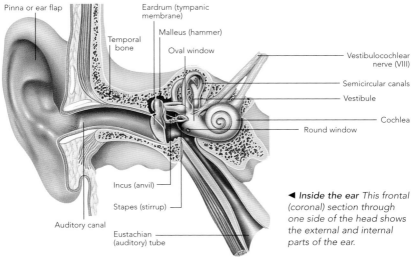

Pinna or ear flap

Eardrum (tympanic membrane)

Temporal bone

Malleus (hammer)

Oval window

Vestibulocochlear nerve (VIII)

Semicircular canals

Vestibule

Cochlea

Round window

Incus (anvil)

Stapes (stirrup)

Auditory canal

Eustachian (auditory) tube

◀ *Inside the ear This frontal (coronal) section through one side of the head shows the external and internal parts of the ear.*

a second membrane-covered opening between inner and middle ears; the **cochlear duct** lies centrally between the other two. Running along the cochlea is the spiral **organ of Corti**. This consists of about 15,000 mechanoreceptors, called hair cells, that 'sit' on the basilar membrane, and from which project 'hairs' or microvilli that touch an overlying, rooflike membrane called the **tectorial membrane**. Nerve fibers from hair cells merge to form the cochlear branch of the vestibulocochlear nerve.

## Hearing

Sound consists of waves of alternating high and low pressure that pass through the air and other media. Humans can detect sounds in the range 20 to 20,000 hertz (Hz), although this range decreases with age. The pinna directs sound waves into the outer ear, causing the eardrum to vibrate. As the eardrum vibrates, the three ossicles move, producing a pistonlike action on the oval window as the base of the stapes is pulled

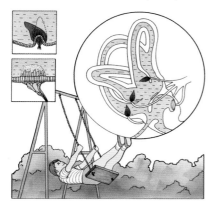

▲ *Balance Ampullae (red) within the three semicircular canals detect movement of the head. Hair cells in the utricle and saccule (yellow) of the vestibule detect whether the head is upright or not. Input to the brain from all these structures enables it to detect the direction the body is moving, and its position relative to gravity. It is then able to control the body's balance and posture by sending instructions to its muscles.*

*The cochlea*
*Cutaway view of the cochlea (arrows show direction of sound waves)*

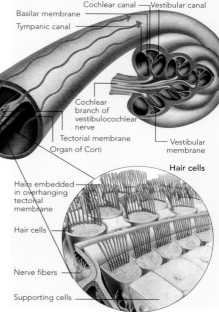

Cochlear canal — Vestibular canal
Basilar membrane
Tympanic canal
Cochlear branch of vestibulocochlear nerve
Tectorial membrane
Organ of Corti
Vestibular membrane

Hair cells

Hairs embedded in overhanging tectorial membrane
Hair cells
Nerve fibers
Supporting cells

in and out. This sets up vibrations in the fluid of the vestibular duct of the cochlea. The vibrations pass through the vestibular membrane, across the cochlear duct, and set up vibrations in the basilar membrane. This causes the hair cells of the organ of Corti to move up and down, thereby distorting their hairs against the tectorial membrane. This bending generates nerve impulses that travel via the thalamus to the temporal lobe of the brain, where they are perceived as sounds. Vibrations returning along the tympanic duct are dissipated at the round window. High-pitched sounds are detected by hair cells near the oval window; low-pitched sounds are detected by hair cells nearer the tip of the cochlea. Loudness depends on the amplitude of the vibrations. The split-second difference in the time taken for sounds to arrive at the left and right ears gives the brain clues as to the direction of the sound.

# ENDOCRINE GLANDS AND HORMONES

The nervous and endocrine (hormonal) systems combine to control and coordinate the body's activities. The nervous system uses high-speed electrical signals that stimulate muscles and glands; it acts rapidly with short-term effects. The endocrine system releases chemicals, known as hormones, that are carried by the blood and regulate the metabolic activities of cells thereby controlling processes such as growth, metabolism, and reproduction; the nervous system works more slowly and has longer-lasting effects. The two systems interact to control body activities.

## Hormones

These chemical messengers are carried to all parts of the body by the bloodstream. Each affects specific target cells. Hormones are classified into one of two groups – amino acid-based hormones and steroid hormones – according to their chemical structure. **Amino acid-based hormones** – including amines, peptides, polypeptides, and proteins – bind to specific receptors in the target cell membrane. This initiates chemical reactions that alter the target cell's metabolism. **Steroid hormones** – the hormones released by the ovaries and testes, and by the cortex of the adrenal glands – penetrate the target cell's membrane and interacts directly with genes in the nucleus to alter metabolism.

## Endocrine glands

Endocrine glands, or ductless glands, release their secretions – hormones – into the bloodstream. The major endocrine glands are the pituitary, pineal, thyroid, parathyroid, thymus, and adrenal glands. Other organs, including the testes, ovaries and pancreas, have significant areas of endocrine tissue and are also considered as endocrine organs. These organs, together with other areas of hormone-producing tissue, make up the endocrine system.

## Pituitary gland

The pituitary gland, the most important of the endocrine glands, releases nine or more hormones that either target other endocrine glands and stimulate them to release hormones, or directly control body functions, such as growth. The hypothalamus controls the operation of the pituitary gland. A description of the structure and function of this gland is on pp. 58–59.

## Pineal gland

Hanging from the roof of the third ventricle in the posterior diencephalon, this tiny gland secretes **melatonin**. Melatonin secretion follows a 24–hour cycle, with levels rising at night and falling around midday, in response to variations in light levels relayed to the pineal by the hypothalamus. High melatonin levels promote drowsiness, and the hormone is believed to help set the body's 'internal clock'.

## Thyroid gland

Situated in the front of the neck below the larynx, the thyroid secretes three hormones. Thyroxine and triiodothyronine act on all body cells, increasing their metabolic rate and the pace of cell division. **Calcitonin** decreases levels of calcium in the blood by reducing the rate at which bone is broken down.

## Parathyroid glands

Embedded in the posterior part of the thyroid gland, the four tiny parathyroid glands secrete parathyroid hormone (PTH) which has the opposite effect to calcitonin because it increases the level of calcium in the blood by stimulating bone breakdown. Together, through negative feedback, PTH and calcitonin maintain stable levels of calcium in the blood.

## Thymus gland

Located posterior to the sternum, the thymus gland produces a number of hormones that are responsible for the normal development of T-lymphocytes and the immune response (see pp. 72–73). It is most active in children, and shrinks throughout adulthood.

## Pancreas

Situated inferior and posterior to the stomach, the pancreas has an endocrine portion, that releases two hormones – insulin and glucagon. **Insulin** reduces blood glucose levels by stimulating cells to take up glucose, and the liver to store glucose as glycogen. **Glucagon** increases blood glucose levels by stimulating the liver to break down glycogen and release glucose into the bloodstream. Insulin and glucagon work together to maintain stable blood glucose levels.

## Adrenal glands

Sitting like 'hats' on top of the kidneys, the two adrenal glands divide into two parts. The **outer cortex** secretes steroid hormones called **corticosteroids**: glucocorticoids help control cell metabolism and reduce stress; and **mineralocorticoids** control levels of sodium and potassium in the blood. The inner **adrenal medulla** secretes **norepinephrine** and **epinephrine** – two hormones that, untypically, work rapidly to help the body deal with stress, and produce the 'fight or flight' reaction.

## Ovaries and testes

As well as producing and releasing sex cells (ova and sperm), the ovaries and testes also release hormones – **estrogen** and **progesterone** by the ovaries, and **testosterone** by the testes – that maintain secondary sexual characteristics and stimulate sex-cell production. Their roles are described in greater detail on pp. 98–101.

## Regulating hormone levels

A process known as **negative feedback** controls hormone levels. If, for example, blood glucose levels rise, this stimulates release of insulin, but inhibits release of glucagon, from the pancreas, causing levels to fall. As glucose levels fall, this inhibits insulin release but stimulates glucagon release, causing glucose levels to rise.

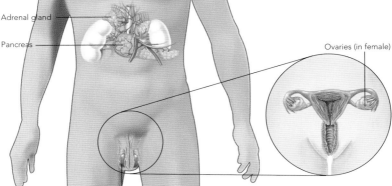

Hypothalamus
Pituitary gland

Parathyroid gland
Thyroid gland

Thymus gland

Adrenal gland

Pancreas

Ovaries (in female)

Testis (in male)

◀ *Endocrine system*
*This diagram shows the position of the body's main endocrine organs.*

# PITUITARY GLAND

The pituitary gland, or hypophysis, lies just below the brain, protected within the **sella turnica** of the sphenoid bone. Despite its relatively small size, the pituitary releases nine or more hormones, many of which trigger other endocrine glands to release hormones of their own. For that reason the pituitary gland is often referred to as the 'maestro of the endocrine orchestra' because it affects the activities of many other endocrine glands. It is, in fact, itself under the control of the **hypothalamus**, a part of the brain to which the pituitary is directly connected (*see* pp. 58–59). Together, the pituitary gland and the hypothalamus provide a direct link between the two coordinating and control systems of the body – the nervous and endocrine systems.

## Structure of the pituitary gland

The pea-sized pituitary gland is attached to the hypothalamus superiorly by a stalk called the **infundibulum**. It consists of two lobes: one (anterior) glandular and the other (posterior) made of neural tissue. The **anterior lobe**, or adenohypophysis, makes up about 70 percent of the pituitary gland. It originates from the oral mucosa and is therefore composed of epithelial tissue. The hypothalamus gland communicates with the anterior lobe by way of the hypophyseal portal veins in the infundibulum. Neurons in the hypothalamus secrete release (and inhibiting)

hormones, commonly polypeptides, that enter the **primary capillary plexus** in the superior infundibulum and are carried by the **hypophyseal portal veins** to the **secondary capillary plexus** in the anterior pituitary, where they regulate secretion of hormones. Together the primary plexus, hypophyseal portal veins, and secondary plexus form the **hypophyseal portal system**; a portal system is an arrangement of blood vessels where, unusually, capillaries merge to form veins which then branch into capillaries.

The **posterior lobe** (neurohypophysis) is part of the brain and is connected to the hypothalamus through a bundle of nerve fibers called the hypothalamic-hypophyseal tract, that runs through the infundibulum. The tract arises from neurons in the hypothalamus called neurosecretory cells. They make two hormones, known as neurohormones because of their mode of production, called

⮕ Thyroid-stimulating hormone

⮕ Growth Hormone

⮕ Prolactin

⮚ Follicle-stimulating hormone

⮕ Luteinizing hormone

⮕ Adrenocortropic hormone

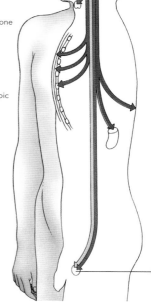

Pituitary gland and hypothalamus

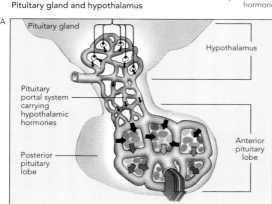

A
Pituitary gland

Hypothalamus

Pituitary portal system carrying hypothalamic hormones

Anterior pituitary lobe

Posterior pituitary lobe

antidiuretic hormone (ADH) and **oxytocin** – which are transported along nerve fibers to the posterior lobe. When hormones are required, the neurosecretory cells 'fire' and hormones are released from the axon terminals into blood capillaries in the posterior lobe and thence to their target tissues.

## Anterior-lobe hormones

The anterior lobe of the pituitary gland synthesizes and releases its hormones in response to releasing hormones received from the hypothalamus. Two of the six anterior lobe hormones – **growth hormone** and **prolactin** – cause a direct effect on target organs; the other four – thyroid-stimulating hormone (TSH), adrenocorticotropic hormone (ACTH), follicle stimulating hormone (FSH) and luteinizing hormone – are referred to as **trophic hormones** because they regulate the action of other endocrine glands. The actions of these six hormones are now summarized.

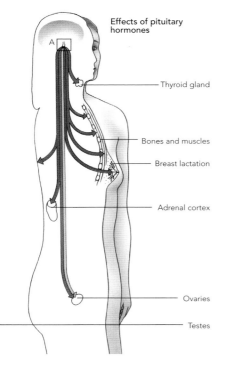

Effects of pituitary hormones

— Thyroid gland

— Bones and muscles

— Breast lactation

— Adrenal cortex

— Ovaries

— Testes

• **Growth hormone (GH)**
Also called somatotropin, GH targets body cells, especially those in bone and muscle and stimulates growth, in children, and repair by stimulating cell division.
• **Thyroid-stimulating hormone (TSH)**
Also called thyrotropin, TSH targets the thyroid gland and stimulates the release of two hormones, thyroxine and tri-iodothyronine that, in turn, accelerate metabolic rate.
• **Adrenocorticotropic hormone (ACTH)**
ACTH targets the cortex (outer part) of the adrenal glands and stimulates the release of corticosteroid hormones that help regulate metabolism and play an important part in helping the body resist stress.
• **Follicle-stimulating hormone (FSH)**
In women, FSH targets the ovaries and stimulates maturation of an ovum and production of the sex hormone estrogen. In men, FSH targets the testes and stimulates sperm production
• **Luteinizing hormone (LH)**
In women, LH targets the ovaries and triggers ovulation, and stimulates the release production of the sex hormones estrogen and progesterone. In men, LH (sometimes called interstitial cell-stimulating hormone or ICSH) stimulates production of the sex hormone testosterone.
• **Prolactin**
Prolactin targets the mammary glands in a women's breasts and stimulates milk production during pregnancy.

## Posterior-lobe hormones

As described above, the posterior lobe receives its hormones from the hypothalamus. Both affect target organs directly, and their actions are now outlined.
• **Antidiuretic hormone (ADH)**
Also called vasopressin, ADH targets the kidneys and increases the amount of water returned to the blood during urine production, thereby reducing urine volume, conserving water and helping the body to maintain its water balance.
• **Oxytocin**
In women, oxytocin targets the uterus and stimulates muscle contraction at birth; it also stimulates the mammary glands to release milk during breast-feeding.

# CIRCULATION

The circulatory or cardiovascular system consists of the blood vessels and the blood that is pumped along them by the heart. The circulatory system plays a vital role in maintaining homeostasis by controlling the concentration and composition of tissue fluid, by supplying cells with essential materials and removing their wastes, by helping to keep the body warm, and by protecting the body against attack by pathogens. A vast network of blood vessels carries blood from the heart to the tissues and back to the heart once again in an unending circulation of this life-giving fluid.

## Twin circulations

There are in fact two circulations within the cardiovascular system. These double, or twin, circulations consist of two loops that are linked by the two sides – left and right – of the heart, which acts as a double pump. The shorter of the two loops, the **pulmonary circulation**, carries oxygen-poor blood from the right ventricle of the heart to the lungs, where it picks up oxygen, and back to the left atrium of the

Anatomy of the circulation

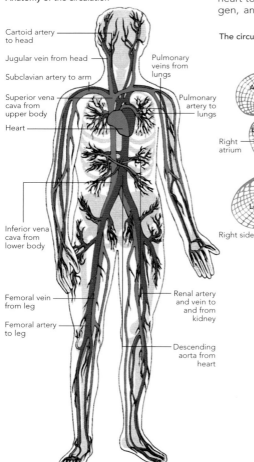

Cartoid artery to head

Jugular vein from head

Subclavian artery to arm

Superior vena cava from upper body

Heart

Pulmonary veins from lungs

Pulmonary artery to lungs

Inferior vena cava from lower body

Femoral vein from leg

Femoral artery to leg

Renal artery and vein to and from kidney

Descending aorta from heart

The circulatory system

Aorta

Pulmonary artery

Heart

Arm

Arm

Lung

Lung

Right atrium

Left atrium

Pulmonary vein

Leg

Leg

Right side

Right ventricle

Left ventricle

Left side

▲ *Schematic circulatory system*
*This diagram shows in schematic form the pulmonary and systemic circulations.*

◄ *Anatomy of the circulation*
*This body map shows the main blood vessels in the body. Arteries (red) usually carry oxygen-rich blood away from the heart. Veins (blue) usually carry oxygen-poor blood toward the heart. The third type of blood vessels, capillaries, link arteries and veins but are too small to be shown here.*

heart. The pulmonary artery, which carries blood to the lungs is unusual in being an artery that caries oxygen-poor blood; the pulmonary veins are similarly unusual because they are veins that carry oxygen-rich blood.

The longer loop, the **systemic circulation**, carries oxygen-rich blood from the heart to the rest of the body, and back to the heart. Oxygen-rich blood leaves the left ventricle of the heart and arrives through numerous branches in the tissues where it unloads oxygen and nutrients, and picks up metabolic wastes and returns along the veins to the right ventricle of the heart. Blood then reenters the pulmonary circuit. A complete journey around the body takes about one minute.

## Portal systems

Most veins carry blood back to the heart. However, in some cases veins carry blood from one organ to another. For example, the hepatic portal (or portal) vein carries food-rich blood from the capillaries of the small intestine to the capillaries of the liver, where the food is processed. A second vein, the hepatic vein, then collects venous blood from the liver and carries it toward the heart. This arrangement is called the hepatic portal system. The hypophyseal portal system carries releasing hormones

▶ The heart pumps blood around the body. The aorta (large red artery, upper left) carries blood to the body. To the left of the aorta is the upper part of the vena cava (blue). This vein carries oxygen-poor blood into the heart from the upper body. This blood then travels from the heart to the lungs (where it obtains oxygen) through the pulmonary artery (blue, center) and returns to the heart through the pulmonary veins.

from the hypothalamus to the anterior lobe of the pituitary gland (*see* pages 58–59).

## The heart

This living pump contracts without a break to pump blood around the circulatory system. It consists mainly of cardiac muscle fibers, that can contract of their own accord and without tiring. The heart has two halves – left and right – separated by a septum. The right side of the heart receives oxygen-poor blood from the body and pumps it along through the pulmonary circulation to the lungs, and then back to the left side of the heart. This pumps oxygen-rich blood through the systemic circulation to the body and then back to the right side of the heart once again. The internal structure of the heart and the mechanism of the heartbeat are described on pages 62–63. Heartrate speeds up during exercise, and slows down when the body is at rest, because of the effects of the two branches of the autonomic nervous system on the pacemaker within the heart that controls its rate of contraction.

## Blood pressure

When the lower chambers of the heart, the ventricles, contract, they exert a high pressure on the walls of the arteries that carry blood away from the heart. This pressure provides the driving force to push the blood around the body. As the ventricles relax before the next contraction, the blood pressure falls. The contraction is called systole, and the relaxation diastole. The higher (systolic) and lower (diastolic) pressures can be measured using an instrument called a sphygmomanometer. This is performed routinely by doctors and nurses to ensure that a patient is healthy, and that their blood-pressure readings are neither too high nor too low.

## Position
The heart lies in the thorax behind the sternum and in front of the descending aorta and esophagus. It rests on the central ligament of the diaphragm muscle. On either side are the two lungs. Above are the main blood vessels and the bifurcation of the trachea into the two main bronchi.

## Basic facts
The heart weighs about 11 ounces (300 grams) and is about the size of a grapefruit. It has two atria, two ventricles, and four valves. It receives blood from the two venae cavae and four pulmonary veins and expels it into the aorta and pulmonary artery. It pumps 19,000 pints (9,000 liters) of blood a day at a rate varying between 60 and 160 beats a minute.

## Anatomy
The heart is surrounded by the fibrous pericardium containing the serous peri-cardial sac holding a small amount of fluid that allows frictionless movement. It consists of two pairs of chambers – atrium and ventricle – that act as separate pumps. The right side pumps deoxy-genated blood through the lungs – the **pulmonary circulation**. The left side pumps oxygenated blood from the lungs – the **systemic circulation**.

## Atria and ventricles
The blood from the superior and inferior venae cavae enters the right atrium. The four pulmonary veins take blood to the left atrium.

The atrioventricular valves, tricuspid in the right and mitral in the left, have special muscles (the papillary muscles) and fine tendons (chordae tendineae) attached to the edges of the valve cusps to prevent them rupturing back into atria during ventricular systole.

The left ventricle has thicker muscle than the right to sustain a higher systemic

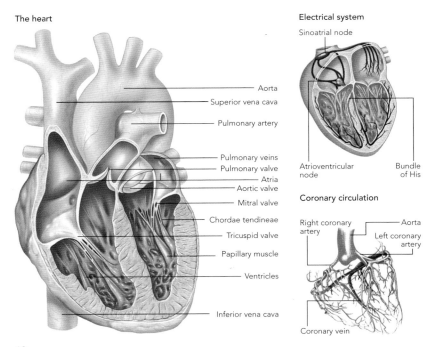

The heart

Aorta
Superior vena cava
Pulmonary artery
Pulmonary veins
Pulmonary valve
Atria
Aortic valve
Mitral valve
Chordae tendineae
Tricuspid valve
Papillary muscle
Ventricles
Inferior vena cava

Electrical system
Sinoatrial node

Atrioventricular node
Bundle of His

Coronary circulation

Right coronary artery
Aorta
Left coronary artery

Coronary vein

blood pressure. The ventricles are closed by the aortic and pulmonary valves.

The heart is lined with endocardium, and divided into halves by the interatrial and interventricular septa.

The remnants of the fetal circulation can still be seen in the fibrous band of the ductus arteriosus.

## The electrical system

To keep the heart beating, the sinoatrial node (or 'pacemaker') in the right atrium sends impulses through the two atria,

causing atrial systole. It then stimulates the atrioventricular node to pass impulses rapidly down the bundle of His to cause ventricular systole (see artwork, near left).

## Coronary circulation

The myocardium has its own blood supply (*bottom left*) from the left and right coronary arteries – the first branches of the aorta – supplying different areas of muscle with a few anastomoses between them. The coronary veins drain venous blood into the right atrium.

▶ *A heartbeat*

*In atrial diastole (A) blood flows from the inferior and superior venae cavae into the right atrium (1) and from the four pulmonary veins into the left atrium (2). The flow is increased during inspiration as the negative intrathoracic pressure also 'sucks' blood into the heart as well as air into the lungs. This results in sinus arrhythmia.*

*When ventricular systole ceases (B), the intraventricular pressure drops, and the two atrioventricular valves – the tricuspid (3) and the mitral (4) – float open and blood starts to flow from the atria (1, 2) into the ventricles. The sinoatrial node then initiates atrial systole and the blood is pumped past the valves, which are fully open, into the dilated ventricles.*

*Atrial systole ceases (C) when the electrical impulse reaches the atrioventricular node and passes down the bundle of His to start ventricular systole. Once atrioventricular valves (3, 4) snap closed, the chordae tendineae and papillary muscles prevent them*

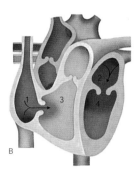
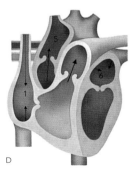
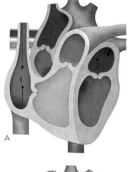
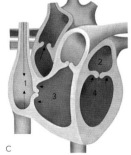

*bursting back into the atria. Venous blood can again flow into the atria (1, 2), during atrial diastole and ventricular systole.*
*The rapidly increasing ventricular pressure (D) throws open the aortic (5) and pulmonary (6) valves with blood streaming into*

*the systemic and pulmonary circulations. The elasticity of the arterial walls causes the valves (5, 6) to snap closed at the end of ventricular systole. The 'snap' of the opening and closing heart valves can be heard as 'lub-dub' through the chest wall.*

# BLOOD

Pumped by the heart, blood flows along blood vessels, supplying the body's trillions of cells with all their needs, while ensuring that they are kept in stable surroundings. Blood makes up about eight percent of an adult's mass, with 8.5 to 10.5 pints (4 to 5 liters) of blood in an woman, and 10.5 to 12.5 pints (5 to 6 liters) in a man. Blood is a connective tissue: red blood cells, white blood cells, and platelets are suspended in its liquid matrix, plasma.

## Functions of blood

Blood underpins homeostasis by helping to maintain a stable internal environment inside the body. The homeostatic functions of blood can be divided into three basic types: transportation, regulation, and protection. Blood **transports** oxygen from the lungs to all body cells, and glucose, amino acids, and other nutrients from the small intestine. It removes metabolic wastes, such as carbon dioxide from all body cells and urea from the liver, and carries them to a point of elimination from the body, respectively the lungs and the kidneys. Blood helps **regulate** the body's internal environment by maintaining a constant pH in both blood and tissue fluid, and by helping to maintain a constant body temperature by distributing heat from sources of production, such as the liver and muscles, around the body. Blood **protects** the body by preventing leakage from damaged blood vessels through blood clotting, and by carrying white blood cells to sites of infection where they destroy invading pathogens.

## Plasma

Yellow plasma is 90 percent water, seven percent plasma proteins, and three percent dissolved small molecules. Dissolved plasma proteins include albumin, which helps maintain normal blood volume; alpha and beta globulins, which help transport some lipids and hormones; gamma globulins, or antibodies, which defend the body against infection; fibrinogen and other proteins involved in clotting; and certain antibacterial proteins and metabolic enzymes. Other solutes include glucose, amino acids, fatty acids, and other nutrients; carbon dioxide and nitrogenous waste products; hormones; and sodium, potassium and other ions that help to maintain the normal concentration and pH level of the blood.

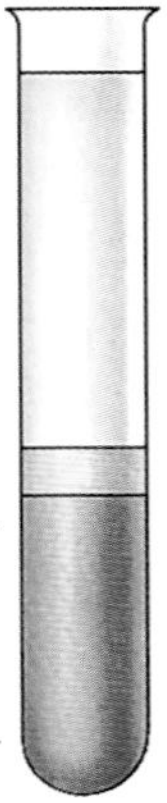

▶ **Blood composition**
If blood is spun in a centrifuge tube, its main constituents separate and their relative volumes can be estimated visually. The heaviest components – red blood cells – sink to the bottom of the tube, while the lightest – plasma – remains at the top. White blood cells and platelets form a thin layer in between. Plasma makes up about 55 percent of the total volume, red blood cells about 44 percent, and white cells and platelets about one percent.

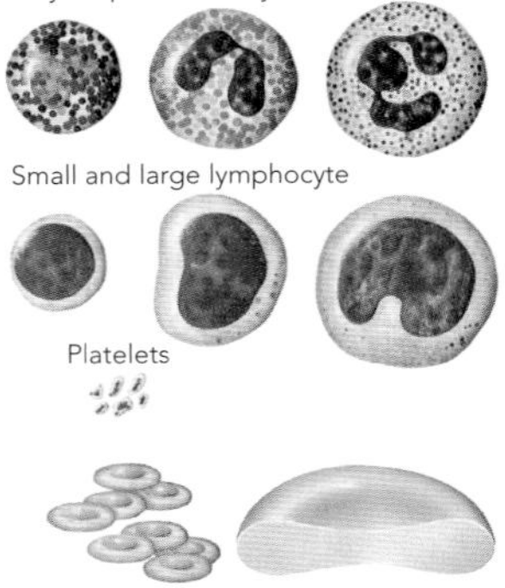

Plasma

White cells
and platelets

Red blood
cells

Polymorphonuclearcytes in immature to mature form

Small and large lymphocyte

Monocyte

Platelets

Red blood
cells

▲ **Blood cells** This illustration shows many of the different types of blood cells (magnified). The three main types of white blood cells are polymorphonuclearcytes, lymphocytes, and monocytes. Platelets are vital clotting agents. Red blood cells are the most numerous.

## Red blood cells

Erythrocytes or red blood cells make up about 99 percent of blood cells, with some 5 million cells per cubic millimeter of blood, outnumbering white blood cells by 800 to one. Their function is to supply oxygen to all body cells. Each red blood cell is a flattened disk with a dimpled center, giving it a biconcave profile in section. This shape provides a large surface area, relative to the cell's volume, for taking up and releasing oxygen. Red blood cells are also small – at about 7 micrometers, just narrower than the diameter of the smallest capillaries – and have a natural elasticity, which enables them to squeeze along narrow capillaries in single file. Red blood cells lack a nucleus; instead they are packed with the protein **hemoglobin**, which gives them their red color. As red blood cells pass through the lungs, where oxygen concentration is high, hemoglobin picks up oxygen to form oxyhemoglobin; when red blood cells pass through the tissues, where oxygen levels are low because of its constant consumption by cells in cell respiration, oxyhemoglobin unloads its oxygen and becomes hemoglobin. Oxyhemoglobin gives arterial blood its bright red color, while hemoglobin gives venous blood its dark red color.

## White blood cells

Although there are far fewer white blood cells than red blood cells – about 7,000 per cubic millimeter of blood – they play a vital role in defending the body against infection by pathogens. White blood cells have the ability to squeeze through capillary walls, by a process known as **diapedesis**, and enter the tissues. White blood cells are divided into three groups: granulocytes, monocytes, and lymphocytes. **Granulocytes** are first on the scene if pathogens enter the body or if tissue is damaged. The commonest types, called **neutrophil**, are active phagocytes: they detect, surround, engulf, and digest pathogens and debris. **Monocytes** are also phagocytic; in the tissues they become large, voracious hunters called **macrophages**. **Lymphocytes** play a key part in the immune response and are described in more detail on pages 72–73.

## Platelets

Platelets are cell fragments that lack a nucleus. They are about one-third the size of red blood cells, and are described in more detail on pages 66–67.

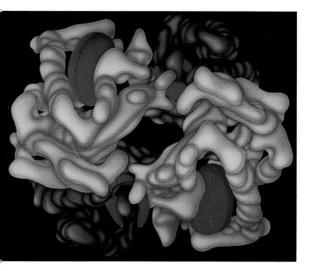

◄ *Hemoglobin*
*This computer-generated model shows the structure of a hemoglobin molecule. It consists of four subunits. Each subunit consists of a polypeptide chain (there are two pairs – blue or orange); and a heme molecule (red disk), a red pigment that contains at its core an iron atom. Each heme group can bind one oxygen molecule, so with 250,000 million molecules packed into a single red blood cell, that cell can carry about one billion oxygen molecules when fully*

## Blood cell formation

Blood cells are produced, in adults, by red bone marrow located within the spongy bone found in the flat bones of the skeleton – skull, collar bones, shoulder blades, sternum, ribs, vertebrae, and hip bones – and in the epiphyses of the femur and humerus. The cells that divide to produce blood cells are **stem cells** called hemocytoblasts. Their derivatives differentiate to form either red blood cells, granulocytes (neutrophils, eosinophils, or basophils), agranulocytes, or platelets.

While granulocytes continue to mature in bone marrow, the precursors of monocytes and lymphocytes move to lymph nodes, spleen, and other organs of the lymphatic system and continue their development there.

Red blood cells are produced, by a process called **erythropoiesis** at the rate of about 2 million per second. They have a life span of about 120 days, at which point they are worn out and need replacing. Old red blood cells are processed in the spleen and liver and their iron content

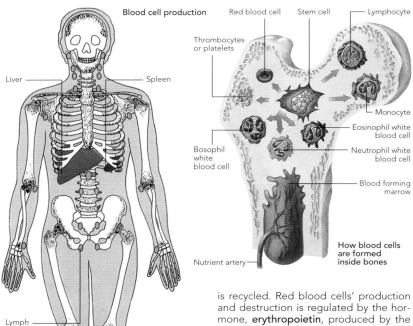

**Blood cell production**

Liver

Spleen

Lymph glands

Sites of blood-cell production in the body

Red blood cell　　Stem cell　　Lymphocyte

Thrombocytes or platelets

Monocyte

Eosinophil white blood cell

Neutrophil white blood cell

Bosophil white blood cell

Blood forming marrow

Nutrient artery

**How blood cells are formed inside bones**

is recycled. Red blood cells' production and destruction is regulated by the hormone, **erythropoietin**, produced by the kidneys. This hormone responds to reductions in blood oxygen levels, produced perhaps by moving to a higher altitude, by increasing the rate of erythropoiesis. Production of white blood cells, known as **leukopoiesis**, is controlled by a number of hormones. Most granulocytes die within days as a result of combating pathogens; monocytes live for several months; whereas lymphocytes, the backbone of the immune system, can live for many decades.

## Blood clotting

Blood clotting prevents excessive blood loss from a wound. Normally, circulating blood contains red cells, platelets, plasma, clotting factors, and fibrinogen. Tissue-clotting factors lie trapped within cells surrounding each blood vessel (A). When damage occurs blood escapes from the broken vessel. Platelets congregate at the site and help plug the wound. Tissue-clotting factors are released (B). The reaction of the platelets with plasma and tissue-clotting factors converts the soluble fibrinogen into insoluble threads of fibrin. The fibrin forms a mesh across the break (C). Platelets and blood cells become trapped in the mesh. The gelatinlike mass shrinks and serum oozes out, leaving a clot (D). The clot dries out to form a scab under which the damaged tissue repairs itself.

## Blood groups

Although the red cells in different people look the same they are, in fact, dissimilar. They can be divided into four main groups, A, B, AB, and O (below). The surface of the cells in each group is different and will act as an antigen to plasma from another group, which carries the antibody. This causes the cells to stick to each other by the process of agglutination. An individual with Group A cells will carry the antibody B in his plasma, those in group AB do not carry either antibody, while those in group O have both antibodies but the cells do not have either antigen. Thus blood from any group can be transferred into those of group AB, as they do not carry antibodies; group AB individuals are known as 'universal recipients.' Group O recipients can only receive

**Blood clotting**

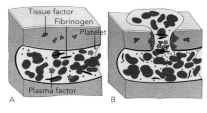
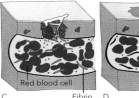

blood from another group O donor, but can give blood to anyone, and thus are known as 'universal donors.'

Antibodies in the plasma of the donor blood are quickly diluted by the recipient and the concentration is therefore too low to cause agglutination. The two commonest groups in western Europe are Groups A and O, each found in about 45 percent of the population, group B in about 10 percent, and group AB is found in less than 5 percent.

The antigen-antibody reaction not only causes agglutination, but also hemolysis – the breakdown of the red blood cells releasing hemoglobin into the circulation. This is an 'incompatible transfusion reaction' and can lead to fever, jaundice, kidney failure from blocking of the tubules with hemoglobin, and even, in some cases, death.

▶ **Blood group compatability**
This chart shows blood group compatability between donor and recipient. Where the red blood cell symbol is smooth, there is compatability, where the symbol is rough, there is incompatibility. So, for example, a group O donor can donate blood to any other group, but can only receive blood from another group O person.

| Recipient | A | B | AB | O |
|-----------|---|---|----|----|
| Donor A | smooth | rough | smooth | rough |
| B | rough | smooth | smooth | rough |
| AB | rough | rough | smooth | rough |
| O | smooth | smooth | smooth | smooth |

# BLOOD VESSELS

Laid end to end, the blood vessels of the body would stretch more than 60,000 miles (100,000 kilometers). The three main types of blood vessels – arteries, veins, and capillaries – are living, flexible structures that maintain blood flow in one direction, respond to the body's changing demands by diverting blood flow to particular areas, and, in the case of capillaries, are 'leaky' enough to allow the outflow and inflow of fluid to deliver materials to, and collect them from, the body's cells.

Apart from capillaries, blood vessels share the same structure (*see* below) with four distinct layers or tunics – **endothelium** which allows the smooth flow of blood; **connective tissue; smooth muscle** and elastic fibers to confer strength and flexibility; and a tough outer coat of **collagen** fibers – surrounding the central lumen. The three types of blood vessels differ in their lengths and diameters and the relative thickness of their walls.

## Arteries

Arteries are elastic, thick-walled vessels that carry blood under high pressure away from the heart. All carry oxygen-rich blood, apart from the pulmonary trunk (artery) which carries oxygen-poor blood from the right ventricle to the lungs. The largest artery, the **aorta**, is about 1 inch (2.5 cm) in diameter, and carries oxygen-rich blood out of the left ventricle. Arteries branch repeatedly as they get further away from the heart, giving rise to progressively smaller vessels. **Elastic arteries** are closest to the heart, and include the aorta. When blood passes along them, their highly elastic walls first expand and then recoil – an action that smoothes out the dramatic pressure changes that occur when the heart contracts and then relaxes. The pressure wave passing along an elastic artery is called a pulse, and can be felt at various points in the body, such as the wrist. **Muscular arteries** are smaller branches of elastic arteries. Contraction of smooth muscle fibers in the walls or muscular arteries, under the control of the autonomic nervous system, enables control of the flow of blood along the arteries. The smallest muscular arteries divide to form arteries, vessels less than 0.01 inches (0.3 mm) in diameter. These divide to form capillaries.

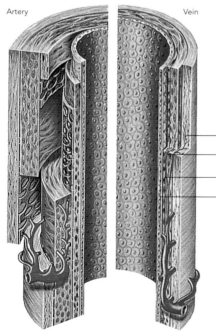

Artery                Vein

Both arteries and veins are tubes made of four layers:

Protective fibrous covering

Smooth muscle and elastic fibers

Connective tissue

Endothelial layer of cells

◄ *Comparing arteries and veins*
*The arteries carry blood away from the heart and the veins return it. Both arteries and veins (right) are made in the same way with four layers: a protective fibrous coat; a middle layer of smooth muscle and elastic tissue, which is thickest in the largest arteries; a thin layer of connective tissue; and a smooth layer of cells – the endothelium.*

## Capillaries

The smallest blood vessels, having a lumen of 8–10 micrometers in diameter, capillaries nonetheless make up some 90 percent of the length of the blood vessel network. Their role is to get blood to all tissue cells. The capillary wall is just one cell thick and is leaky. Fluid, but not cells, from the blood passes out through the wall and into the tissue fluid that bathes cells. This is how oxygen, food, and so on are transferred from blood to cells. As the capillaries weave their way through the tissues in capillary beds, the osmotic pressure of the blood increases, and this draws fluid containing wastes and secretions from cells back into the blood. Capillaries then merge to form **venules**, the smallest vein branches.

## Veins

Veins carry blood from the tissues back to the heart. All carry oxygen-poor blood, except for the pulmonary vein which carries oxygen-rich blood from the lungs to the left atrium of the heart. The venules unite to form larger veins that finally empty into the inferior vena cava (lower body) and superior vena cava (upper body) that drain into the right atrium of the heart. Veins have thinner walls than arteries, as the blood that passes though them is under much lower pressure. (*See* venous blood flow below)

▲ *This scanning electron micrograph (SEM) image shows red blood cells (erythrocytes) passing from a capillary to a vein.*

### Veins and venous blood flow

*The blood reaches the venous system through minute capillary vessels. It is through the capillary wall that oxygen and carbon dioxide, food, and metabolites are exchanged with the interstitial fluid. Most of the interstitial fluid returns to the venous system, but some is collected by the lymphatic vessels.*

*The returning venous blood moves slowly due to low pressure and the veins can collapse or expand to accommodate variations in blood flow (right). Movement relies on the surrounding muscles, which contract (1) and compress the vein. Pulsation of adjacent arteries (2) has a regular pumping effect.*

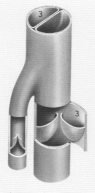

*Semilunar valves (3) are found at regular intervals throughout the larger veins and these allow the blood to move only in one direction. They are more common in the legs. The veins frequently anastomose with each other so that the blood flow can*

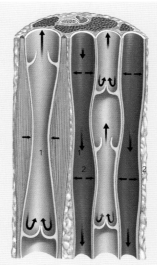

*alter direction if there is any constriction or pressure from movement of muscles or ligaments.*

# LYMPH AND LYMPH ORGANS

In addition to the cardiovascular system, the body has a second, less familiar transportation system – the lymphatic system. It consists of a network of blind-ending tubes, called lymphatic, or lymph, vessels and a number of associated lymphatic organs. The lymphatic system has two main functions: the transport of a fluid called lymph, and the provision of sites for the cells of the immune system (see pages 72–73) to attack and destroy invading pathogens.

## Lymph

The lymphatic system plays a vital role in maintaining the normal volume of the blood. Every 24 hours some 50 pints (24 liters) of fluid passes through the walls of blood capillaries to supply food and oxygen to tissue cells. It then removes carbon dioxide and other materials and returns through the capillary walls into the blood. However, some 6 to 8 pints (3 to 4 liters) does not return in this way, and

would accumulate in the tissues if it were not collected by fine, blind-ending lymph capillaries. These tiny vessels have tiny flaps that act like one-way swing doors, which allow fluid into the lymph capillaries but not back into the tissues. The watery fluid, now called lymph, is returned by the lymph vessels to the bloodstream in order to ensure that its volume remains unchanged.

## Lymph vessels

Reaching nearly every part of the body, lymph vessels form a one-way network of channels. Fine lymph capillaries permeate the tissues, where they collect lymph. They join to form larger vessels, called lymph-collecting vessels, that merge to form larger vessels called lymph trunks that empty into the two largest lymph vessels, the thoracic duct and right lymphatic duct, that empty lymph into, respectively, the left subclavian vein and right subclavian vein and thereby back into the bloodstream. This completes the journey made by the fluid 'left behind' in the tissues back to the bloodstream. The lymphatic system has no equivalent of the heart to push lymph along them.

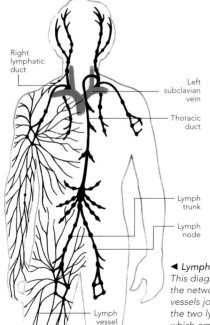

Right lymphatic duct

Left subclavian vein

Thoracic duct

Lymph trunk

Lymph node

Lymph vessel

◀ *Lymph vessels*
*This diagram shows how the network of lymph vessels join up to form the two lymphatic ducts, which empty lymph into the bloodstream.*

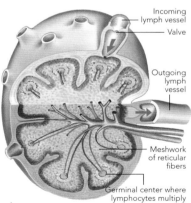

Incoming lymph vessel

Valve

Outgoing lymph vessel

Meshwork of reticular fibers

Germinal center where lymphocytes multiply

▲ *Lymph node*
*This section though a lymph node shows the direction of lymph flow, and the meshwork of fibers that support macrophages and lymphocytes.*

### Spleen

*Situated to the left of the stomach, and served by the splenic artery and veins, the spleen is the largest of the lymph organs. It has two main roles. First, it removes aging red blood cells from the blood and salvages their iron for later use. Second, its macrophages remove pathogens and cell debris from the blood flowing though its sinuses (blood spaces), while lymphocytes initiate the immune response. White pulp consists of fibers carrying lymphocytes, while red pulp consists of blood sinuses and areas of connective tissue rich in macrophages.*

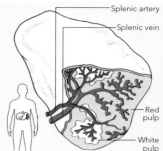

Splenic artery
Splenic vein
Red pulp
White pulp

Instead this function is achieved by the skeletal muscles that surround lymph vessels and push lymph along them when they contract. Like veins, lymph vessels also have valves that prevent the backflow of lymph.

## Lymph organs

Lymph (or lymphoid) organs including lymph nodes, the spleen (*see below*), and tonsils, form the part of the lymphatic system that defends the body by destroying pathogens and cancer cells. They contain the white blood cells' lymphocytes and macrophages that form a vital part of the immune system.

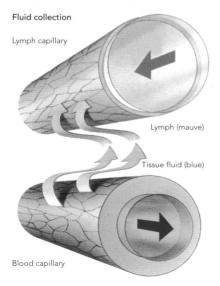

Fluid collection

Lymph capillary

Lymph (mauve)

Tissue fluid (blue)

Blood capillary

## Lymph nodes

These small bean-shaped swellings, each between 1–20 mm in diameter, occur along lymph vessels. The lymph that passes through them contains not just clear tissue fluid but also bacteria and other pathogens, cell debris, and possibly cancer cells. Inside a lymph node is a meshwork of fibers that carries large numbers of macrophages and lymphocytes; lymph nodes also produce large numbers of lymphocytes, up to ten billion per day. As lymph passes though the lymph node, macrophages and lymphocytes detect and destroy pathogens and cancer cells, and engulf any cell debris. The 'filtered' lymph then continues on its way back to the bloodstream. During certain infections the lymph nodes become inflamed and tender, a condition known as 'swollen glands.'

## Tonsils

Two tonsils at the back of the tongue, two at the back of the mouth, and one in the upper throat, form a ring of lymph tissue that protects the entrance to both digestive and respiratory systems. Lying in the mucosa, each tonsil contains blind-ending crypts that contain lymphocytes. These destroy bacteria, carried into the body by food or air, that migrate into the crypts.

## Peyer's patches

These are small patches of lymph tissue found in the wall of the lower small intestine and the appendix. They contain macrophages and lymphocytes that prevent the bacteria that are commonplace in the intestines from entering the human bloodstream.

71

# DEFENSES AGAINST INFECTION

Every second, the body is exposed to a range of pathogens – including bacteria, viruses, protists, and fungi – that can, if they enter the body's tissues and bloodstream, multiply and destabilize the body's normal homeostasis, and thereby cause infections that may be fatal. Nonspecific defenses repel and attack any pathogens, while the specific defenses – the immune system – destroy specific pathogens and retain a memory of them.

## Nonspecific defenses

These are built into the body from birth, and provide two lines of defense: surface barriers, and defensive cells and chemicals.

## Barriers and secretions

Surface barriers provide the first line of defense against pathogens and include the tightly-packed cells of the skin, and those of the mucous membrane lining the digestive, respiratory, urinary, and reproductive systems. Secretions include **mucus** secreted by the mucous membranes of the nasal cavity, trachea, and

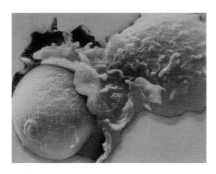

▲ *Phagocytosis*
*This SEM shows a phagocyte engulfing a pathogen. The phagocyte recognizes the bacterium, or other pathogen, by the chemical signals it sends out. It wraps itself around and engulfs the pathogen. The pathogen is taken into the phagocyte and digested. Neutrophils are often destroyed by this and form a mass of white liquid called pus at the site of infection. The process of binding phagocyte to pathogen is made more effective by coating the pathogen with complement.*

bronchi. Mucus traps pathogens, is carried to the throat, swallowed, and digested by acidic gastric juice in the stomach. **Gastric juice** destroys bacteria entering the body in food or drink. Tears contain antibacterial lysozyme which kills bacteria as the tears flow over the eye.

## Cellular and chemical defenses

Should pathogens penetrate epithelial barriers and invade the tissues, a second line of nonspecific defenses involving defensive cells and chemicals swings into action.

• **Phagocytosis**
Phagocytes, or 'cell eaters,' are white blood cells (*see* pages 64–65) that are attracted to sites of infection or damage, and include neutrophils and macrophages that trap and engulf prey by phagocytosis (*see* illustration, below left).
• **Natural killer (NK) cells**
Found in lymph nodes and circulating in the blood, NK cells target cells infected by viruses as well as cancer cells. NK cells bind to these rogue cells and destroy them with toxic chemicals.
• **Inflammation**
When tissue damage occurs, local and white blood cells release chemicals including histamines that send out an alarm signal, making local blood vessels wider and more leaky, bringing in extra phagocytes and antibodies to fight infection. Increased blood flow also makes the area warm, swollen, and tender, resulting in inflammation.
• **Antimicrobial proteins**
Two main groups of specific defense proteins are produced. **Interferon** is released by cells already infected by viruses. It binds to nearby cells and protects them from viral infection. **Complement** is a system of proteins that, during infection, enhances the immune response and inflammation. Complement also opsonizes pathogens – makes them more 'tasty' – so that they become easier targets for phagocytes.
• **Fever**
The abnormally high temperature that occurs with some infections serves to inhibit bacterial multiplication and speed up the rate of repair by body tissues.

## Specific defenses: the immune system

The immune system consists of macrophages and lymphocytes found particularly in the lymphatic, but also in the cardiovascular systems. It produces a highly targeted and specific defense system that is both adaptive and flexible. It is triggered by markers called **antigens** that are carried by pathogens and abnormal normal body cells. Immune system cells retain a memory of antigens, so that should the same pathogen be encountered again, the immune system launches a rapid and highly effective strike to destroy it. The immune system has two parts: the humoral (or antibody-mediated) immune response, controlled by **B-lymphocytes**; and the cell-mediated immune response, controlled by **T-lymphocytes**. Both types of lymphocytes are produced in bone marrow, but B-lymphocytes mature in bone marrow, while T-lymphocytes mature in the thymus gland. As lymphocytes mature, or become immunocompetent, each becomes capable of responding to a specific antigen. The mature lymphocytes then disperse to lymph nodes, the spleen, and other lymph tissue. When the immune system is confronted by a pathogen for the first time, it takes several days to respond to it – this is the primary response – during which symptoms of illness may appear. Should the same pathogen invade again later, the immune system responds immediately. This explains why people rarely suffer the same disease twice because they have become immune to it.

## Humoral immune response – B lymphocytes

When a pathogen, such as a bacterium, enters the body it is engulfed by a macrophage, which 'presents' the pathogen's antigen to a matching B lymphocyte in the lymph system that is primed to recognize that specific antigen. The B lymphocytes divide rapidly to produce cells that spread to the blood and tissues and release chemicals called antibodies that bind to the pathogen and mark it for destruction by phagocytes. Division of B lymphocytes also produces memory B-cells that remember the pathogen and respond strongly should it ever return.

## Cell-mediated immune response: – T lymphocytes

This part of the immune system responds well to cells invaded by viruses and cancer cells – both have altered antigens. Macrophages engulfing these cells 'present' antigens to T lymphocytes. Lymphocytes multiply rapidly, producing both killer T lymphocytes that lock onto their target and kill it with chemicals; and memory T cells that retain a memory of the antigen for future reference.

▼ *At the centre of lymphocyte* cells are large nuclei with yellow chromatin.

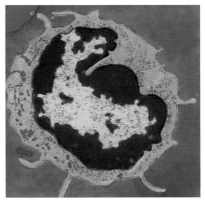

▲ B lymphocyte          ▼ T lymphocyte

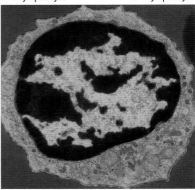

# AIRWAYS

Although the body can survive without nutrition or water for a period of time, it has an absolute requirement for a continuous supply of oxygen. The reasons for this are twofold. First, oxygen is needed by each one of the body's trillions of cells in order to release the energy that is locked up in glucose and other fuel molecules. This energy is needed to power the many reactions going on inside cells that allow them to operate as living units. Second, oxygen cannot be stored by the body. Energy is released from glucose, with the involvement of oxygen, in the mitochondria of cells by cell respiration (see pages 6–7), a process that releases carbon dioxide (and water) as a waste product.

The source of oxygen is the air around the body. By volume, oxygen makes up some 20 percent of the atmosphere at sea level. However, it is unfeasible to absorb oxygen through the skin's surface and supply it to all body cells. Instead the respiratory system takes air into the body, and facilitates the entry of oxygen into the bloodstream which carries it to all body cells. At the same time, the respiratory system removes carbon dioxide – which would poison the body if allowed to accumulate – and expels it from the body into the atmosphere.

The respiratory system consists of the lungs (see pages 76–77) – through which oxygen enters, and carbon dioxide leaves – the bloodstream; and the airways – the nose, pharynx (throat), larynx, trachea, and bronchi – that carry the air between the lungs and the atmosphere, and are described below. Air is sucked into, and pushed out of, the lungs by the action of breathing.

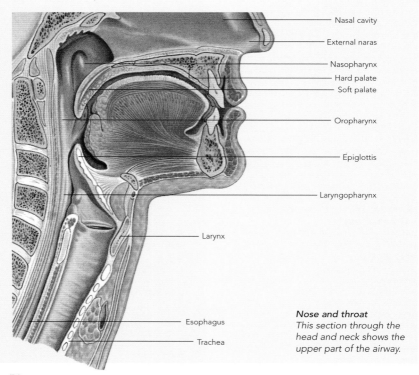

Nasal cavity

External naras

Nasopharynx

Hard palate

Soft palate

Oropharynx

Epiglottis

Laryngopharynx

Larynx

Esophagus

Trachea

**Nose and throat**
*This section through the head and neck shows the upper part of the airway.*

## Nose

While air can be breathed in through either the mouth or nose, intake through the nose ensures that the air is filtered, moistened, and warmed before it contacts the delicate tissues found inside the lungs. The nose consists of the external visible part, supported by cartilage, and the inner nasal cavity that is divided into left and right halves by the nasal septum. On the lateral side of each half there are three projecting 'ledges' formed by bones called the **conchae**, or turbinates. Like most of the airways, the nasal cavity is lined with ciliated mucous membrane that is well supplied with blood and secretes watery mucus. When air is breathed in through the external nares, or nostrils, the coarse hairs that guard the entrance to the nasal cavity remove larger dirt and dust particles. As the air passes over the conchae, it becomes turbulent and swirls over the nasal cavity lining. Heat from thin-walled blood vessels, and moisture from the watery mucus warms and moistens the air. At the same time, dust and other particles become trapped in sticky mucus, which is wafted by the hairlike cilia to the throat and swallowed, and digested by acidic gastric juice in the stomach which also kills any bacteria. Thus cleaned, air passes into the pharynx.

## Pharynx

Looking like a short length of garden hose, the pharynx or throat, links the nose to the larynx. It provides a pathway for not only air, but also for food passing to the esophagus. The pharynx has three continuous sections: the **nasopharynx** runs from the internal nares to the soft palate; the **oropharynx** connects with the back of the throat; and the **laryngopharynx** divides into two parts: the anterior larynx and the posterior esophagus.

## Larynx

The funnel-shaped larynx, or voice box is about 2 inches (5 cm) long and links the pharynx to the trachea. Its structure and role in sound production is described on pages 80–81. The **epiglottis**, the leaf-shaped flap of cartilage that is hinged at the superior, anterior edge of the larynx, folds posteriorly over the entrance to the larynx during swallowing (*see* pages 84–85) to prevent food going 'the wrong way' toward the lungs.

## Trachea

The trachea, or windpipe, is about 11 cm (4 inches) long and runs behind the sternum to link the larynx to the lungs. It is reinforced by up to 20 C-shaped pieces of cartilage that prevents it from collapsing when air is sucked into the lungs. The open part of the 'C' occurs where the trachea adjoins the esophagus, which runs parallel and posterior to the trachea. The tracheal epithelium secretes mucus which traps dust, while cilia propel the dirt-laden mucus to the throat where it is swallowed. At its inferior end, the trachea branches into two bronchi which enter the lungs and branch repeatedly (*see* pages 76–77).

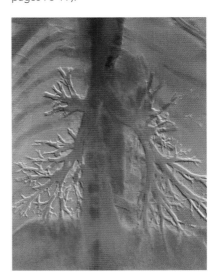

▲ *Bronchial tree*
*This false-color bronchograph clearly shows the bronchial tree in a human lung. The airways in the lungs resemble an upside-down tree in which the trachea is the trunk, the bronchi are the main branches, and the smaller bronchi and bronchioles form the branches and twigs.*

# LUNGS AND GAS EXCHANGE

While the rest of the respiratory system is concerned with conveying air in and out of the body, the function of the lungs is to get oxygen into, and carbon dioxide out of, the bloodstream. Roughly cone shaped, the two lungs fill most of the thoracic cavity apart from the central mediastinum, which contains the heart, major blood vessels, trachea, bronchi, and esophagus. The apex of each lung extends above the clavicle, while its base rests on the diaphragm, the sheet of muscle that separates the thorax from the abdomen, and which plays a key role in breathing (see pages 78–79). The lungs are surrounded and protected by the rib cage, formed by the sternum and ribs,

and the thoracic section of the vertebral column. Surrounding each lung is a double pleural membrane, or pleura: the **outer pleura** lines the inside of the thoracic cavity and the superior surface of the diaphragm; the **inner pleura** covers the lungs. Pleural fluid secreted by the pleural membranes ensures that during breathing the lungs slide painlessly over the thoracic wall. Each lung is divided into lobes by surface fissures: the right lung has two fissures and three lobes; the left lung has one fissure and two lobes, and is smaller than its partner because of an indentation in its medial surface called the cardiac notch, which provides room for the heart.

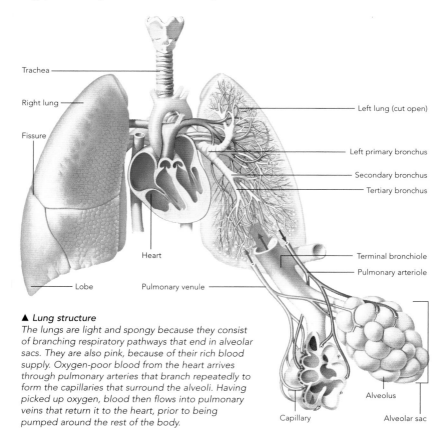

Trachea

Right lung

Fissure

Left lung (cut open)

Left primary bronchus

Secondary bronchus

Tertiary bronchus

Heart

Terminal bronchiole

Pulmonary arteriole

Lobe

Pulmonary venule

Alveolus

Capillary

Alveolar sac

▲ *Lung structure*
*The lungs are light and spongy because they consist of branching respiratory pathways that end in alveolar sacs. They are also pink, because of their rich blood supply. Oxygen-poor blood from the heart arrives through pulmonary arteries that branch repeatedly to form the capillaries that surround the alveoli. Having picked up oxygen, blood then flows into pulmonary veins that return it to the heart, prior to being pumped around the rest of the body.*

## Bronchi and bronchioles

As already described on pages 74–75, the trachea divides at its inferior end into two bronchi which each enter the lungs at an indentation called the **hilus**, which is also the entry and exit point for the pulmonary blood supply. Inside the lungs, each bronchus divides to form secondary bronchi. These divide to form tertiary bronchi, which in turn split into smaller bronchi and even smaller bronchioles that are no more than 0.04 inches (1 mm) wide. These further subdivide into terminal bronchioles, about 0.02 inches (0.5 mm) wide, of which there are about 30,000 in each lung. Terminal bronchioles subdivide into two or more respiratory bronchioles that terminate in alveolar sacs, which resemble bunches of grapes. Each sac consists of a number of alveoli, and it is through these that oxygen and carbon dioxide are exchanged.

## Alveoli and gas exchange

The 300 million alveoli in both lungs provide a surface area for gas exchange of about 750 square feet (70 square meters), or 35 times the surface area of the skin, crammed into a space no larger than a shopping bag. This large surface ensures that sufficient oxygen can be taken into the body to satisfy the demands of 100 trillion cells. Each alveolus is covered by a network of pulmonary blood capillaries. Its inner surface is covered by a thin film of fluid that contains a surfactant which prevents the alveoli from collapsing. The wall of the alveolus and that of the blood capillary adjacent to it forms the respiratory membrane. Oxygen diffuses across the membrane down a concentration gradient – high oxygen in alveolus, low oxygen in blood – from alveolus to blood. Carbon dioxide moves in the opposite direction – carbon dioxide high in blood, low in alveolus. The constant inflow of oxygen-rich air into the lungs, and the removal of oxygen-rich blood from the lungs, maintains these concentration gradients. The distance across the respiration membrane from the inside of the alveolus to the inside of the blood capillary is only 0.001 mm, ensuring that diffusion can happen rapidly. Diffusion is aided by the fluid layer of the alveolus in which oxygen dissolves prior to crossing the respiratory membrane. Once in the blood, oxygen is carried by red blood cells (see pages 64–65) to the tissues where oxygen passes into cells and waste carbon dioxide moves in the opposite direction.

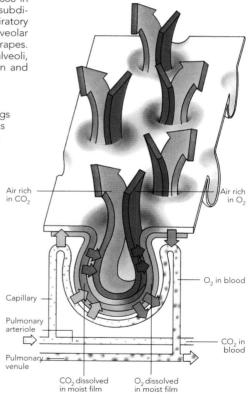

Air rich in $CO_2$

Air rich in $O_2$

$O_2$ in blood

Capillary

Pulmonary arteriole

$CO_2$ in blood

Pulmonary venule

$CO_2$ dissolved in moist film

$O_2$ dissolved in moist film

▲ **Gas exchange in the alveolus** This diagram shows how carbon dioxide diffuses out of the blood and into the alveoli, while oxygen diffuses out of the alveoli and into the blood.

# BREATHING

Air is constantly moved in and out of the lungs by a process called breathing (or ventilation). The normal breathing rate at rest is between 12 and 18 times a minute, although this rate increases during exercise to satisfy the increased oxygen needs of muscles. Apart from their bronchi, air sacs, and blood vessels, lungs consist of large elastic connective tissue, and do not possess the muscles that could perform ventilation. Instead, their innate elasticity enables the lungs to follow the movement of the thoracic cavity in order to draw in or expel air. The lungs are effectively sealed inside the thorax by the thorax wall and diaphragm; the only opening is through the bronchi and trachea. The pleural fluid found between the paired pleural membranes lining the thorax and covering the lungs has an adhesive function, ensuring that as the thorax gets larger or smaller it pulls or pushes the elastic lungs so they too change in volume. The prime mover of breathing is the diaphragm, with the intercostal muscles between the ribs playing an assisting function. When the external intercostals contract, they raise the rib cage upward and outward. When their antagonists the internal intercostals contract, they pull the rib cage downward and inward.

## Inhalation and exhalation

During inhalation (or inspiration), air is breathed in. The dome-shaped diaphragm contracts and pushes the underlying abdominal organs downward; the external intercostal muscles contract and elevate the rib cage. As a result of these two actions, the volume inside the thorax increases, and so – as described by Boyle's Law – the pressure inside the thorax decreases. Since the lungs follow the movements of the thorax, the volume of the lungs increases and their pressure decreases below that of the atmosphere outside the body. This negative pressure-differential causes air to be sucked into the lungs. During quiet breathing, such as when someone is sitting at a desk, the diaphragm does most of the work, with the external intercostals only coming into play when that person becomes more active, and their rate and depth of breathing increases.

During exhalation, or expiration, air is breathed out. The diaphragm relaxes and is pushed upward into a dome shape by the underlying abdominal organs. The ribs move downward and inward. Both actions cause the volume of the thoracic cavity to decrease, and the naturally elastic lungs follow suit. The pressure inside the lungs increases above atmospheric pressure, and air is forced out. During quiet breathing, once again, most of the work is done by the diaphragm. During forced exhalation, when a person is exercising vigorously, the internal intercostals come into play by actively returning the rib cage to its downward position.

## Lung volumes

The lungs never completely empty; if they did, they would collapse and it would prove difficult to reinflate them. This means that the lungs are not completely flushed out and refilled with 'fresh' air with each breath. Instead, breathing refreshes air remaining in the lungs by adding more oxygen and removing carbon dioxide. The volume of air inhaled and exhaled with each breath depends on the

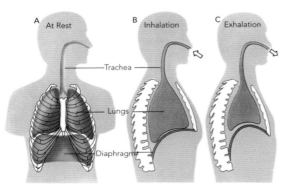

A  At Rest    B  Inhalation    C  Exhalation

Trachea

Lungs

Diaphragm

▲ *Inhalation and exhalation The diaphragm contracts during inhalation and relaxes during exhalation.*

age and sex of the individual and their fitness. Doctors and other health specialists use certain respiratory volumes (see chart below) to assess their patients' health. Respiratory volumes are measured using a spirometer, an instrument that measures the volume of air passing in and out through a mouthpiece.

## Control of breathing

The respiratory center, in the medulla oblongata of the brain, automatically controls the breathing muscles through the phrenic nerve to the diaphragm and through the many intercostal nerves to the intercostal muscles. It can also make use of the accessory muscles of respiration when they are required.

The respiratory center has two sources of information about breathing: the degree to which the lungs are stretched and the level of carbon dioxide in the bloodstream.

The stretching of the lungs through inspiration is detected by branches of the vagus nerve, which reflexly inhibit inspiration and allow expiration to take place. The fluctuations in the level of carbon dioxide are detected by nerve endings in the aorta and carotid arteries. An increase in carbon dioxide causes an increase in acidity of the blood and this stimulates the respiratory rate – a condition known as hyperpnea.

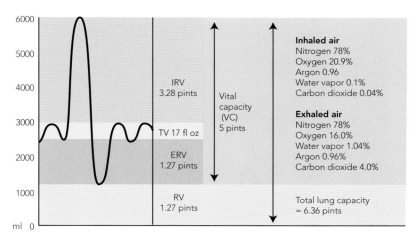

▲ Lung volumes
Respiratory volumes and respiratory capacities – combinations of particular respiratory volumes – are used to assess a person's respiratory health. The figures given here apply to a healthy male.
• Tidal volume (TV) is the volume of air that is inhaled and exhaled with each breath during quiet breathing.
• Inspiratory reserve

volume (IRV) is the volume of air that can be forcibly inhaled in addition to TV.
• Expiratory reserve volume (ERV) is the volume of air that can be forcibly exhaled in addition to TV.
• Residual volume (RV) is the air that remains in the lungs after the most forceful exhalation.
• Inspiratory capacity (IC) is the sum of TV ≠ IRV, and is the total amount of air that can be inhaled forcibly.

• Functional reserve capacity (FRC) is the sum of RV + ERV and is the amount of air remaining in the lungs after exhalation during quiet breathing.
• Vital capacity (VC) is the sum of TV + IRV + ERV and is the total volume of air that can be inhaled and exhaled.
• Total lung capacity is the sum of IRV + TV + ERV + RV. It is the total volume of air in fully inflated lungs.

# SPEECH AND BREATHING

The feature that distinguishes humans from other animals is the ability to communicate using speech sounds produced by the controlled release of exhaled breath though the vocal cords in the larynx, and modified by tongue and lips. In addition, other respiratory movements produce sounds such as coughing and hiccuping.

## Larynx and vocal cords

The larynx, or voice box, is situated superior to the trachea, and can be felt as a firm projection in the anterior neck. It consists of nine pieces of cartilage, and associated muscles, held together by ligaments. These pieces of cartilage are the ring-shaped **cricoid cartilage** that attaches the larynx to the trachea; the **thyroid** and the paired **arytenoid cartilages** that anchor the vocal cords, the folds of fibrous tissue that stretch from the anterior to the posterior sides of the larynx; the **epiglottis**, the leaf-shaped cartilage flap that folds over the larynx and prevents food going 'the wrong way' toward the lungs during swallowing; and the paired **cuneiform** and **corniculate cartilages** that comprise part of the lateral and posterior walls of the larynx. In men, the anterior prominence of the thyroid cartilage can be felt in the neck as the so-called 'Adam's apple.'

There are two pairs of vocal cords. The upper pair, the **vestibular folds** or false vocal cords, do not play a part in sound production but help prevent food being inhaled. The lower pairs of true vocal cords are responsible for sound production. Between the true vocal cords is an opening called the **glottis**. This stays open during normal breathing.

## Sound and speech production

Sounds are produced by the larynx during exhalation. Laryngeal muscles attached to the arytenoid cartilages contract to pull the vocal cords together. The forcing of exhaled air between the closed vocal cords forces them to vibrate, setting up sound waves in the air in the pharynx, nasal cavity, and the mouth. The more taut the vocal cords are pulled by the laryngeal muscles, the faster the vibrations (and shorter the wavelength of the sound waves) and the higher-pitched the sounds. The slacker the vocal cords are pulled and the wider the glottis, the more low-pitched the sounds. Men naturally have lower-pitched voices than women because their vocal cords are thicker (this change occurs at puberty) and so vibrate more slowly. The volume of sound depends on the force with which air is sent between the vocal cords. During shouting, for example, air is sent through the glottis

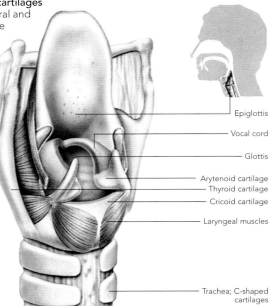

Epiglottis

Vocal cord

Glottis

Arytenoid cartilage
Thyroid cartilage
Cricoid cartilage

Laryngeal muscles

Trachea; C-shaped cartilages

▶ *Section though the larynx*
*This vertical section through the larynx shows its main cartilage plates and the laryngeal muscles.*

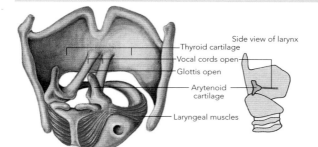

Side view of larynx

Thyroid cartilage
Vocal cords open
Glottis open
Arytenoid cartilage
Laryngeal muscles

◀ *Sound production*
*These two diagrams show how the movement of the arytenoid cartilages by the laryngeal muscles close the glottis by pulling the vocal cords together and making them taut.*

Posterior views of larynx

Side view of larynx

Vocal cords closed
Arytenoid cartilage
Cricoid cartilage

with tremendous force and at high speed, producing sound waves of high amplitude with a resulting loud sound.

Raw sounds produced during breathing are then modified to produce the recognizable sounds in speech. The pharynx, nasal cavities, and paranasal sinuses act as resonating chambers that amplify the sounds. Movements and shape-changes of the mouth cavity, tongue, and lips produce recognizable vowels and consonants. Vowels are formed by altering the shape of the lips. Consonants are generated by a movement of the tip of the tongue to and from the back of the teeth or roof of the mouth, or by a rapid separation of the lips.

Speech requires the coordinated contraction of a number of different muscles including those that control breathing, the opening and closing of the vocal cords, and the movement of the lips and tongue. Control of these muscles resides in Broca's (speech) area in the left cerebral hemisphere of the brain. Broca's area receives input from Wernicke's area about 'what is to be spoken.' Wernicke's area, in turn, receives input from other parts of the cerebrum including from the auditory cortex, concerning spoken language, and from the visual cortex about written language.

## Other breathing movements

Several other processes, apart from speech, may serve to modify the normal breathing pattern.

• **Coughing** is a reflex action that clears mucus and dust from the airways. A deep breath precedes the closing of the glottis. Muscles increase pressure in the thoracic cavity, and air pressure builds up behind the vocal cords. These open suddenly, and air rushes up the bronchi, trachea, and pharynx, and out of the mouth, clearing the obstruction as it does so.

• **Sneezing** is similar to coughing except that the blast of air is directed out through the nasal cavity from which it removes irritating particles and mucus.

• **Hiccuping** is caused by an irritation of the diaphragm, perhaps by eating too quickly. The diaphragm contracts spasmodically, sucking air into the lungs in short bursts, and making the vocal cords snap shut to produce the characteristic 'hiccup' sound.

• **Yawning** is probably stimulated by excess levels of carbon dioxide in the blood. It involves a deep inhalation with the mouth wide open, which 'flushes out' stale air from the lungs.

81

# FOOD AND DIGESTION

Food – or, to be more specific, the nutrients contained in food – is essential for normal cell function and, therefore, to keep the body alive. Humans are driven to eat by appetite and hunger, both controlled by the hypothalamus in the brain. The various nutrients contained in the food we eat provide cells with energy needed to power metabolism (see p. 8), and the raw materials required for growth and repair. Most foods contain a mixture of nutrients, although these nutrients are often 'locked' inside large molecules that cannot be used by the body because they are too big to be absorbed into the bloodstream. For absorption to occur, food has first to be digested by the digestive system, which works like an assembly line in reverse, converting complex nutrients into simpler ones. The process has four stages.

• **Ingestion** is the process of taking food or drink into the body through the mouth.
• **Digestion** is breaking down of complex nutrient molecules – mainly carbohydrates, fats, and proteins – into simpler ones.
• **Absorption** is the movement of the products of digestion, such as glucose or amino acids, from the alimentary canal into the bloodstream.
• **Egestion** is the elimination of indigestible material, such as fiber, as well as dead cells, and bacteria from the body in the form of feces.

## Digestive system

The digestive system consists of two linked parts: the alimentary canal and the accessory digestive organs. The **alimentary canal** is essentially a tube, some 30 feet (9 meters) long, that extends from the mouth to the anus, with its longest section – the intestines – packed into the abdominal cavity. The lining of the alimentary canal is continuous with the skin, so technically its cavity lies outside the body. The alimentary 'tube' consists of linked organs that each play their own part in digestion: mouth, pharynx, esophagus, stomach, small intestine, and large intestine. The **accessory digestive organs** consist of the teeth and tongue in the mouth; and the salivary glands, liver, gall bladder, and pancreas, which are all linked by ducts to the alimentary canal. The structure and function of the alimentary canal and its accessory organs are described in greater detail on pages 84–93.

## Digestion

As already explained, digestion is the breaking down of complex nutrients, such as fats, starches, and proteins, into simple substances that can be absorbed by the body. The two forms of digestion – mechanical and chemical – work alongside each other. **Mechanical digestion** includes the cutting and grinding action of teeth, and the churning action of the stomach's muscular walls. These actions physically break food up into smaller pieces, so that when food arrives in the small intestine – the main site of chemical digestion – it is in the form of a souplike liquid called **chyme**. **Chemical digestion** involves the action of **enzymes** (see also pp. 8–9). These are biological catalysts that speed up reactions both inside and outside cells. Digestive enzymes are extracellular, working in the lumen of the alimentary canal. They accelerate by thousands or millions of times the breakdown of complex nutrients. Most enzymic activity takes place in the small intestine, although it also occurs in the mouth and the stomach.

## Nutrients

Most food contains a mixture of carbohydrates, fats, proteins, vitamins, minerals, and water. Vitamins and minerals – known as micronutrients because they are needed in small amounts – are typically absorbed unchanged in the small intestine, as is water. The other macronutrients usually need to be digested, unless they are ingested in simple form. Plant foods also contain fiber, consisting of cellulose and other indigestible elements of the plant's skeletal structure. Fiber is an essential element of the human diet because it adds bulk to food and gives gut wall muscles something to act against, speeding the movement of waste along the large intestine.

• **Carbohydrates**
Carbohydrates are important sources of energy. They divide into two groups:

polysaccharides (complex carbohydrates) that do not taste sweet, and **sugars** (monosaccharides and disaccharides) that do. Complex carbohydrates, such as starch, are made up of long chains of glucose molecules. Initially, during digestion, starch is broken down into a disaccharide (two-sugar unit) called maltose, which is then broken down into individual glucose (monosaccharide) molecules. Other di-saccharides, such as sucrose (cane sugar) and lactose (milk sugar) taken in with food, are broken down into their constituent simple sugars, glucose and fructose, and glucose and galactose, respectively. These simple sugars are then absorbed.

• **Proteins**

Proteins perform a wide range of metabolic roles inside cells and are built up from building blocks called amino acids, which are absorbed following digestion. The proteins in food consists of long chains of amino acids, of which there are 20 types. Initially proteins are 'cut' into shorter lengths called peptides, a process that begins in the stomach. Then, in the small intestine, peptidases snip individual amino acids off the end of the peptide chains. The amino acids are then absorbed.

• **Fats and oils**

Also called lipids, fats and oils are required to help build cell membranes, insulate the body, and to provide an energy store. Fats and oils consist of molecules called **triglycerides**, each made up of glycerol and three fatty acids. **Lipases** break down each triglyceride into two fatty acids and monoglycerides (glycerol attached to one fatty acid). These are then absorbed.

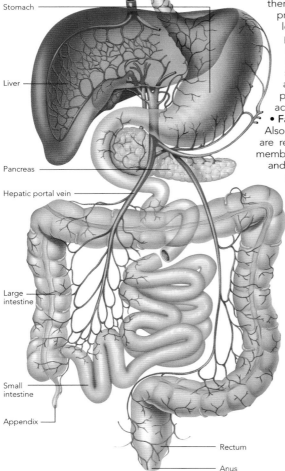
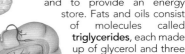

Lower end of esophagus

Stomach

Liver

Pancreas

Hepatic portal vein

Large intestine

Small intestine

Appendix

Rectum

Anus

◄ *Digestive system*
*This diagram shows the main parts of the digestive system. Note that the mouth, pharynx, and most of the esophagus are not shown (see pp. 84–85).*

# CHEWING AND SWALLOWING

## Mouth and chewing

The mouth provides the entrance to both digestive systems. It is bounded by the lips, hard and soft palates, cheeks, and tongue. Posteriorly, the mouth opens into the pharynx. The **lips** are flexible, muscular and highly sensitive flaps that guard the entrance to the mouth. The muscular **cheeks** extend from below the eye to the margin of the jaw. The hard **palate** forms the roof of the mouth, and is supported by bone; the soft **palate** is supported by muscle. The **tongue** is a muscular flap that occupies most of the mouth cavity when it is closed. Intrinsic muscles within the tongue change its shape, while extrinsic muscles, anchored to bone, protrude and retract the tongue. The tongue moves and mixes food during chewing, and houses taste buds (see pp. 48–49) that detect tastes. The **teeth** provide a 'tool kit' of four types of teeth for grasping and breaking up food: chisel-shaped **incisors** grip food and cut it up into manageable pieces; pointed **canines** grip and pierce food; broad-crowned **premolar** teeth crush and chew food; large, broad-crowned teeth, the **molars**, at the back of the jaw, crush food with great force.

Food is ingested by the pulling action of the lips, and the gripping and tearing of the incisors and canines. Once food is inside the mouth, the lips close and the cheeks flatten, as the buccinator muscles contract. The masseter and temporalis (jaw) muscles open and close the jaws, enabling the premolar and molar teeth to crush and grind the food. At the same time, the sight, smell and presence of food in the mouth causes the reflex release into the mouth cavity of saliva from the salivary glands. Watery saliva contains **amylase**, an enzyme that breaks down starch to the disaccharide sugar maltose; mucus which helps bind food particles together and lubricate their passage during swallowing; and **lysozyme**, an antibacterial substance. As the teeth crush food, the tongue mixes the food, then pushes it against the hard palate to form a slippery package called a bolus, and then pushes it backward toward the throat.

## Swallowing

Swallowing has three stages.

• **The oral stage**, already described, occurs when the tongue pushes the food bolus toward the throat. It is under voluntary control, while the rest of the process is automatic, controlled by the autonomic nervous system and a swallowing center in the brain.

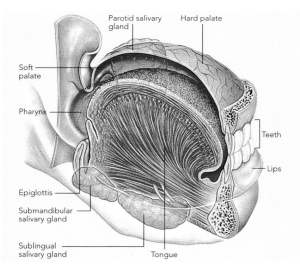

Soft palate

Pharynx

Epiglottis

Submandibular salivary gland

Sublingual salivary gland

Parotid salivary gland

Hard palate

Teeth

Lips

Tongue

*The mouth*
*This diagram shows the main features of the mouth: the tongue, cheeks, hard and soft palates; as well as the epiglottis and pharynx. Three pairs of salivary glands empty about 3 pints (1.5 liters) of saliva through ducts into the mouth daily. These are the parotid glands in front of the ear; the sublingual glands just below the tongue; and the submandibular glands just inside the mandible.*

• **The pharyngeal stage** is triggered when the bolus touches the back of the throat. A wave of muscular contractions pushes the bolus into the esophagus; breathing ceases temporarily; the epiglottis folds over the entrance to the larynx to stop food going toward the lungs.

• **The esophageal stage** pushes food down the esophagus by a wave of muscle contraction called peristalsis. Like much of the rest of the digestive system, the wall of the esophagus contains both circular (inner) and longitudinal (outer) smooth muscle fibers. Circular muscle behind the bolus contracts, while that in front of the bolus relaxes, so food is pushed downward; the waves of contraction and relaxation passes down the esophagus. In addition, longitudinal muscles in front of the bolus contract, making the esophagus wider and shorter. The journey time from pharynx to stomach is between 4 and 8 seconds.

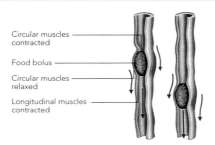

Circular muscles contracted

Food bolus

Circular muscles relaxed

Longitudinal muscles contracted

**◄ ▲ Peristalsis**
*These longitudinal sections through the esophagus outline how the automatic, rhythmic muscular contraction of peristalsis pushes food from the throat to the stomach, as well as moving food in other parts of the digestive system.*

Esophagus

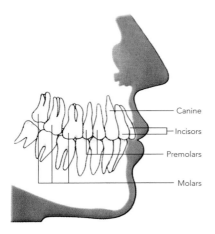

Canine

Incisors

Premolars

Molars

**▲ Permanent teeth**
*Humans have two sets of teeth in their lifetime. The first set of 24 milk, or deciduous, teeth first appear at the age of six months and persist until late childhood. They are gradually pushed out and replaced by the larger set of 32 adult, or permanent, teeth that develop as the jaws get longer. In each half of each jaw there are two incisors, one canine, two premolars, and three molars.*

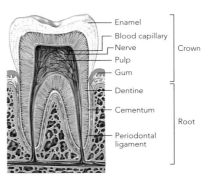

Enamel

Blood capillary

Nerve

Pulp — Crown

Gum

Dentine

Cementum — Root

Periodontal ligament

**▲ Structure of a tooth**
*The teeth are anchored in sockets in the upper and lower jaws and surrounded by the gums. The upper exposed part of the tooth is its crown; the lower part, its root. As this section shows, the crown is covered by enamel, the hardest material in the body. Beneath this is bonelike dentine which shapes the tooth and extends downward into the roots. The inner, soft pulp cavity contains blood vessels and nerves; it produces dentine and provides sensations during chewing. The roots are hold in place by cementum and the periodontal ligament within the tooth socket.*

# STOMACH

The second stage of digestion takes place in the stomach. This J-shaped organ lies between, and connects, the lower end of the esophagus and the duodenum, the first section of the small intestine. The stomach forms the widest, most elastic, and most muscular part of the alimentary canal. This reflects its two main functions. First, it crushes food and churns it up with powerful gastric juices secreted by glands in its lining. Second, it provides a holding area that stores food for three to four hours and then releases it at a slow and steady rate as the small intestine becomes ready to process it. Without this storage area, food would be unable to move along the small intestine to be digested and absorbed efficiently.

## Stomach structure

On average, an adult stomach is about 10 inches (25 cm) long, but, thanks to its elasticity, its volume can vary greatly. When completely empty, the stomach is no bigger than a fist, with a volume of about 1.7 fluid ounces (50 milliliters); its inner lining is thrown into deep folds called rugae. When completely full, after a large meal, its volume can increase to up to 8.5 pints (4 liters), the rugae disappearing as the elastic stomach walls stretch.

The stomach has four regions: the **cardiac region** surrounds the opening of the esophagus; the **fundus** is the upper, dome-shaped section; the **body** is the middle region; while the **pylorus** is the funnel-shaped final section, linked to the duodenum by a ring of muscle called the pyloric sphincter.

The mucous membrane lining the inside of the stomach contains millions of deep gastric pits that contain gastric glands. These glands produce the components of gastric juice: **hydrochloric acid**, necessary to produce the harsh acid conditions inside the stomach; **pepsinogen**, which is turned into the active protein-digesting enzyme by hydrochloric acid; **intrinsic factor**, which is necessary for the absorption of the essential vitamin $B^{12}$ by the small intestine; and, in children, the enzyme **rennin** which converts soluble milk protein into a curd that can be more easily digested. In addition, the glands also produce the hormone **gastrin** which is involved in coordinating digestion.

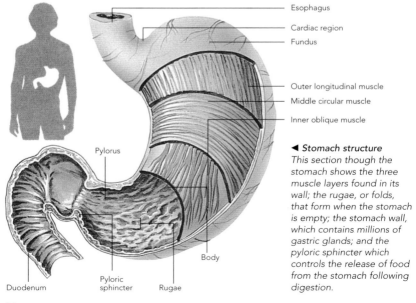

Esophagus

Cardiac region

Fundus

Outer longitudinal muscle

Middle circular muscle

Inner oblique muscle

Pylorus

Duodenum

Pyloric sphincter

Rugae

Body

◀ *Stomach structure*
*This section though the stomach shows the three muscle layers found in its wall; the rugae, or folds, that form when the stomach is empty; the stomach wall, which contains millions of gastric glands; and the pyloric sphincter which controls the release of food from the stomach following digestion.*

Gastric glands also secrete bicarbonate-rich (alkaline) mucus which covers the stomach's lining and protects it from self-digesting by pepsin, or attack by hydrochloric acid. Lining cells are replaced every few days as they become worn by exposure to gastric juice.

In addition to layers of longitudinal and circular muscle, the stomach wall also contains an additional inner layer of oblique muscle. These muscle layers contract in turn to churn up food.

## Digestion in the stomach

Some 4.5 pints (2 liters) of gastric juice are secreted each day inside the stomach. Its hydrochloric acid produces an acidic environment of between pH 1.5 and 3.5, ideal for the action of pepsin (most enzymes work best at a neutral or slightly acid pH) which breaks down proteins to shorter chains of amino acids called poly-peptides. Hydrochloric acid also destroy most bacteria that enter the stomach in food or drink. The release of gastric juice is coordinated to ensure that it is at its peak when food is about to, or has, entered the stomach. There are three overlapping phases of gastric secretion, as outlined below.

• **Cephalic phase** is the release of gastric juice under the control of the autonomic nervous system (ANS) triggered when the brain perceives the thought, sight, smell, or taste of food.

• **Gastric phase** occurs when food arrives in the stomach. The stretching of the stomach wall causes the release of the hormone gastrin which targets the gastric glands that produce it and causes them to release more gastric juice.

• **Intestinal phase** is triggered by arrival of food in the duodenum. This inhibits the release of gastric juice so that the small intestine has time to process the current batch of food before the next one arrives from the stomach.

As food is pushed through the stomach by peristalsis, the waves of contraction become stronger and stronger until they reach the pylorus where they are powerful enough to crush and churn the food. As the ANS stimulates the release of gastric juice, so it increases gastric motility – the speed and strength of contractions of the stomach wall. The crushing of food and its mixing with gastric juice eventually produces a souplike liquid called chyme. This is pushed against the closed pyloric sphincter, which opens slightly to allow squirts of chyme into the duodenum. Any larger, as yet untreated pieces of food are unable to pass through the small opening, snd are recycled for further digestion. The stomach generally empties between two and four hours after a meal, longer if the meal has been very fatty.

**Filling**

Muscular contractions mix food with gastric juice as it enters the stomach.

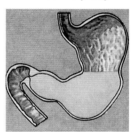

**Digestion**

Strong contractions churn food as gastric juice digests it into chyme.

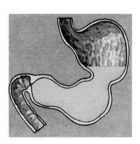

**Emptying**

The pyloric sphincter relaxes to allow liquid chyme into the duodenum.

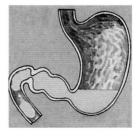

▲ *Gastric filling and emptying*
*This diagram shows the three phases of digestion in the stomach.*

# SMALL INTESTINE AND PANCREAS

The intestines make up about 80 percent of the length of the alimentary canal. They have two sections, and are coiled and folded to fit into the limited space inside the abdominal cavity. The small intestine carries out the major part of food digestion, as well as the absorption of the simple molecules that are the products of digestion. The large intestine plays no part in digestion but serves to absorb water and salts and convey undigested waste out of the body. Both are lined with slippery mucous membrane that secretes lubricating and protective mucus, and both have layers of longitudinal and circular muscle in their walls that facilitate the movement of food or waste along their length by peristaltic and other movements. The structure of the small intestine, and that of the pancreas and gall bladder, both accessory digestive organs, are described here. The digestion and absorption of food by the small intestine, and the structure and functions of the large intestine, are described on pages 90–91.

## Small intestine

The small intestine is the longest and most important part of the alimentary canal. It is between 20–23 feet (6 and 7 meters) long in its relaxed state and about 1 inch (2.5 cm) in diameter. It extends from the pyloric sphincter, which guards the exit from the stomach, to the ileocecal sphincter, which guards the entrance to the large intestine. It has three parts: the duodenum, jejunum, and ileum. The **duodenum** is 10 inches (25 cm) long and receives partially digested food, called chyme from the stomach; bile, a product of the liver, from the gall bladder

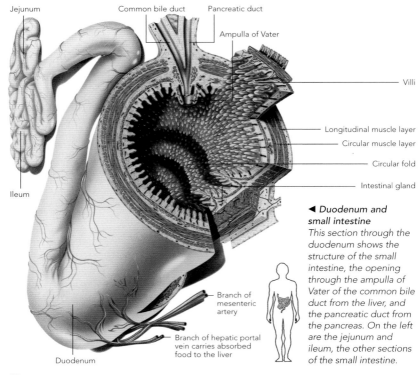

Jejunum

Common bile duct

Pancreatic duct

Ampulla of Vater

Villi

Longitudinal muscle layer

Circular muscle layer

Circular fold

Intestinal gland

Ileum

Branch of mesenteric artery

Branch of hepatic portal vein carries absorbed food to the liver

Duodenum

◀ *Duodenum and small intestine*
*This section through the duodenum shows the structure of the small intestine, the opening through the ampulla of Vater of the common bile duct from the liver, and the pancreatic duct from the pancreas. On the left are the jejunum and ileum, the other sections of the small intestine.*

▼ *Pancreas*
*This section through the pancreas shows both exocrine and endocrine regions.*

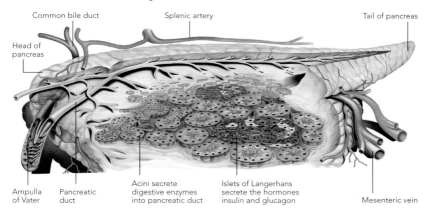

Common bile duct | Splenic artery | Tail of pancreas

Head of pancreas

Ampulla of Vater | Pancreatic duct | Acini secrete digestive enzymes into pancreatic duct | Islets of Langerhans secrete the hormones insulin and glucagon | Mesenteric vein

along the common bile duct; and enzyme-containing pancreatic juice from the pancreas. The common bile duct and pancreatic duct merge just before entering the small intestine, so their secretions flow through a common opening. The **jejunum** is about 8 feet (2.5 meters) long, and links the duodenum to the **ileum**, which is about 12 feet (3.6 meters) long and is where most absorption of nutrients takes place.

The effectiveness of the small intestine in both digestion and absorption is increased by three structural features: circular folds, villi, and microvilli. **Circular folds** (or plicae circularis) extend around the circumference of the small intestine, increasing its surface area and slowing the flow of chyme, so giving more time for digestion. **Villi** are tiny fingerlike processes about 0.04 inches (1 mm) long. They significantly increase the surface area through which food can be absorbed. **Microvilli** are microscopic folds that cover the outer surface of intestinal cells – including those covering the villi – and carry digestive enzymes – called brush border enzymes – that complete the final stages of digestion. These three features give the small intestine a surface area of more than 2,150 sq feet (200 sq m), equivalent to the floor area of an average two-story house.

## Pancreas

This long organ lies almost horizontally below the stomach. Exocrine cells, arranged in clusters called acini, secrete pancreatic juice which passes along the pancreatic duct and empties into the duodenum through the ampulla of Vater. The 3 pints (1.5 liters) of pancreatic juice produced daily contains a number of enzymes. It is alkaline and helps to neutralize the acidic chyme arriving from the stomach, thereby helping to provide an optimal pH for the action of pancreatic and intestinal enzymes. The endocrine portion of the pancreas, the **Islets of Langerhans**, secrete insulin and glucagon, two hormones that control glucose levels in blood (*see* pp. 56–57).

## Gall bladder

About 4 inches (10 cm) in length, this muscular sac is located on the posterior surface of the liver. It stores and concentrates bile, a greenish-yellow liquid, of which about 2.1 pints (1 liter) is produced by the liver daily. Bile contains water, cholesterol, bile pigments (the waste product of the breakdown of hemoglobin from red blood cells), and bile salts that play a part in fat digestion. Bile is carried from the liver to the gall bladder by the common hepatic duct.

# INTESTINES

## Digestion in the small intestine

The digestion of food in the small intestine utilizes enzymes from two sources – pancreatic juice secreted by the pancreas, and the brush border enzymes attached to the microvilli that project from the epithelial cells lining the small intestine. Bile from the gall bladder does not contain enzymes but still has a key role in digestion.

When chyme arrives in the duodenum from the stomach, it triggers the release of two hormones – cholecystokinin (CCK) and secretin – by the intestinal wall. These hormones stimulate pancreatic juice secretion. Pancreatic juice travels along the pancreatic duct and empties into the duodenum through the ampulla of Vater. It contains a number of enzymes, although its protein-digesting proteases are not activated until they reach the small intestine. This prevents the self-digestion of the pancreas. CCK also stimulates release of bile from the gall bladder, and it is squirted into the duodenum at the same time as pancreatic juice and through the same opening. The bile salts in bile emulsify fat globules in food, breaking them up into tiny droplets. These droplets present a very large surface area for pancreatic lipase, the enzyme that digests fats, to act on. Pancreatic juice is alkaline, as is the watery intestinal juice that is secreted by glands in the small intestine wall. This helps to neutralize acidic chyme and provide an optimal, slightly alkaline pH for the activity of both pancreatic and brush border enzymes. The action of these enzymes is shown in the table opposite.

## Absorption by the small intestine

Digestion by brush border enzymes yields molecules that are small enough to be absorbed by villi, the fingerlike projections that cover the small intestine, especially those in the ileum. Glucose and other simple sugars, amino acids, and nucleic acid breakdown products are carried through the cells covering the villi, pass into the blood capillaries inside them, and are carried by the hepatic portal vein to the liver for processing (see pp. 92–93). Fatty acids and monoglycerides (a combination of one fatty acid molecule

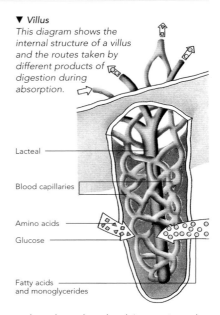

▼ *Villus*
*This diagram shows the internal structure of a villus and the routes taken by different products of digestion during absorption.*

Lacteal

Blood capillaries

Amino acids

Glucose

Fatty acids and monoglycerides

and a glycerol molecule) pass into the epithelial cells of villi, where there are converted back into triglycerides (oils), and then into the lacteals, tiny branches of the lymphatic system (see pp. 70–71). The lymphatic system carries triglycerides away from the ileum and deposits them in the bloodstream. Minerals and vitamins are also absorbed, as is about 95 percent of the 19 pints (9 liters) of water that enters the small intestine each day.

Food spends between 3 to 6 hours in the small intestine. Localized contractions of circular muscles in the intestinal wall moves chyme backward and forward – a process called segmentation – which optimizes both digestion and absorption. The rhythmic contraction of peristalsis is also seen, especially when digestion and absorption are completed and undigested matter is pushed toward the large intestine.

## Large intestine

The large intestine is about 5 inches (1.5 meters) long and 2.5 inches (6.5 cm) in diameter, and extends from the ileocecal valve to the anus. It is divided up into four parts: the cecum, colon, rectum, and anal canal. The pouchlike cecum lies just inferior to the ileocecal valve that maintains a one-way flow from small to large intestines. The colon – the main part of the large intestine – has four regions: the ascending colon passes up the right-hand side of the abdominal cavity; the transverse colon passes across the abdominal cavity just inferior to the liver; the descending colon passes down the right-hand side of the abdominal cavity; and the S-shaped sigmoid colon joins the rectum. The longitudinal muscle of the colon is reduced to three strips, called taeniae, that pull the colon into pockets called haustra. The rectum is about 8 inches (20 cm) long and leads to the 1.2 inches (3 cm) long anal canal which opens to the outside through the anus.

The large intestine does not have a digestive role. It processes the material arriving from the small intestine – undigested food, bile pigments, dead cells – and turns it into semisolid feces that are eliminated from the body. It does this by absorbing water and salts. The large intestine also contains billions of bacteria that feed on undigested material and release some B and K vitamins that are absorbed across the intestinal wall.

Some 3 pints (1.5 liters) of waste arrives each day in the colon from the small intestine. Most of the time, muscles in the intestinal wall contract sluggishly to push waste material slowly along the colon so that water and salts can be absorbed and the waste converted into feces. About three times a day, after meals, a more powerful peristaltic wave – triggered by the arrival of food in the stomach – pushes feces into the sigmoid colon where they are stored until defecation occurs. Every 1 pint (500 ml) of waste entering the colon produces about 5 ounces (150 grams) of feces. The journey from ileocecal calve to sigmoid colon takes between 12 and 36 hours, depending on the type of food.

The arrival of feces in the rectum stretches its wall. This triggers a reflex action, controlled by the autonomic nervous system, that pushes the feces into the anal canal and relaxes the internal and external sphincters that guard the anal opening. At the same time a person has a conscious awareness of the need to defecate. The muscles of the rectum then contract to push the feces out. However, the defecation reflex can be overridden by conscious intervention, and the external sphincter is closed.

▼ *Table: Action of pancreatic and intestinal enzymes in the small intestine. Nutrients in* **bold italics** *are those that are absorbed by the small intestine.*

| Secretion | Enzyme | Action |
|---|---|---|
| Bile from the liver | — | Bile salts emulsify fats into small droplets |
| Pancreatic juice | Amylase | Breaks down starch into maltose from the pancreas |
| | Trypsin | Breaks down proteins into peptides |
| | | Converts inactive chymotrypsinogen into active enzyme chymotrypsin |
| | Chymotrypsin | Breaks down proteins into peptides |
| | Lipase | Breaks down fats and oils droplets into *fatty acids* and *monoglycerides* |
| | Carboxypeptidase | Breaks down peptides into *amino acids* |
| | Nucleases | Break down DNA and RNA into nucleotides |
| Brush border enzymes on intestinal epithelial cells | Enterokinase | Converts inactive trypsinogen into active enzyme trypsin |
| | Peptidases | Break down peptides into *amino acids* |
| | Nucleosidases | Break down nucleotides into *nitrogenous bases, pentoses,* and *phosphates* |
| | Maltase | Breaks down maltose into *glucose* |
| | Sucrase | Breaks down sucrose into *glucose* and *fructose* |
| | Lactase | Breaks down lactose into *glucose* and *galactose* |

# LIVER AND METABOLISM

Weighing 3.3 pounds (1.5 kg), the liver is the body's largest internal organ. The wedge-shaped liver occupies much of the top right-hand side of the abdominal cavity, immediately under the diaphragm. Its deep red color is indicative of the large volume of blood that passes through it. The liver's 500 metabolic and regulatory functions play a key part in homeostasis by ensuring that the blood's composition remains constant. While the liver's only direct role in digestion is to produce bile, many of its functions involve processing the products of digestion after they arrive from the small intestine.

## Liver structure

The basic units of the liver are the billions of **hepatocytes**, or liver cells, that process nutrients, produce bile, destroy poisons, and secrete substances. They are arranged not at random but into highly organized, microscopic structures called lobules. Each **lobule** consists of vertical plates of hepatocytes that radiate from a central core (*see* below). At each of the six corners of a lobule there are three vessels: a branch of the hepatic portal veins that delivers nutrient-rich blood; a branch of

the hepatic artery that delivers oxygen-rich blood; and a branch of the bile duct. Blood from both blood vessels mixes inside blood spaces called **sinusoids** that run between the plates of hepatocytes. It then empties into the central vein that runs along the core of the lobule before emptying into a hepatic vein. As blood flows along the sinusoid, hepatocytes remove oxygen, as well as nutrients and other substance to be processed, and add secretory products, nutrients released form store, and waste products. Sinusoids contain macrophages, called **Kupffer cells**, that engulf worn-out red blood cells, debris, and bacteria. Bile produced by the hepatocytes flows along the bile **canaliculi** that run between plates of hepatocytes and collects in a branch of the bile duct. It is then carried by the bile duct to be stored in the gall bladder.

## Major functions of the liver

By carrying out their 500 or so metabolic functions, the hepatocytes control the chemical composition of the blood. The main functions are described below.

• Blood glucose regulation

Cells need a constant supply of glucose to provide energy to drive their metabolic reactions. So it is essential that glucose levels stay relatively constant. If blood glucose levels are high, after a meal, the liver sequesters glucose and stores it in the form of the polysaccharide glycogen. If levels are too low, between meals, the liver converts glycogen to glucose, which is released

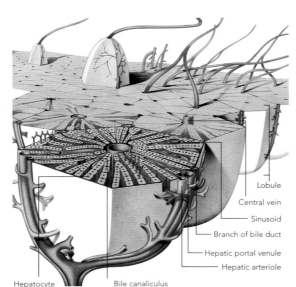

Lobule

Central vein

Sinusoid

Branch of bile duct

Hepatic portal venule

Hepatic arteriole

Hepatocyte          Bile canaliculus

◄ *Liver lobules*
*This magnified view inside the liver shows the hexagonal lobules that make up its structure. Each consists of radiating sheets of hepatocytes between which run the blood-carrying sinusoids and bile-carrying canaliculi.*

into the blood. This regulatory process also involves the mediation of the pancreatic hormones insulin and glucagon (see pages 56–57).

• **Fat metabolism**

If liver glycogen stores are full, the liver converts excess glucose to fat, which can be stored either in the liver or in adipose tissue. The liver also produces cholesterol, a precursor of some steroid hormones and a component of the outer plasma membrane of cells.

• **Mineral storage**

Iron and copper, both needed to make the hemoglobin inside red blood cells, are stored inside hepatocytes.

• **Vitamin storage**

Vitamins A, D, and $B_{12}$ are among the vitamins stored by the liver. The liver can hold up to four months' supply of vitamins D and $B_{12}$, and about two years' supply of vitamin A.

• **Protein metabolism**

Excess amino acids, the building blocks of proteins, cannot be stored in the body. The liver processes the excess by deaminating the amino acids to form urea, a nitrogen-containing waste which is carried to the kidneys to be excreted in urine. In addition, the liver produces plasma proteins and fibrinogen found in blood (see pages 64–65).

• **Bile production**

Bile consists of bile pigments derived from the breakdown of red blood cells; bile salts derived from cholesterol; cholesterol itself; and fats. Bile salts play a part in fat digestion in the small intestine and are then absorbed back into the bloodstream, returned to the liver, and reused. The other bile components are excreted with the feces.

• **Hormone breakdown**

The liver removes hormones from the blood and breaks them down.

• **Detoxification**

Poisonous substances, such as alcohol, that have been inhaled or ingested are detoxified by the liver, which converts them into harmless substances that can be excreted by the kidneys.

• **Heat generation**

The many metabolic reactions going on inside the hepatocytes generates a considerable amount of heat. This is distributed around the body by the blood and helps maintain a constant body temperature.

▶ *Liver's blood supply*
*The liver has two blood supplies: 20 percent is provided by the hepatic artery which supplies oxygen-rich blood; 80 percent arrives along the hepatic portal vein that delivers blood poor in oxygen but rich in nutrients, and other materials, from the small intestine. A portal system is one that carries blood from one organ to another rather than returning it to the heart. The hepatic artery and hepatic portal vein enter the liver midway down its posterior surface. Inside the liver, the oxygen-rich and oxygen-poor/nutrient-rich blood is mixed. 'Processed'*

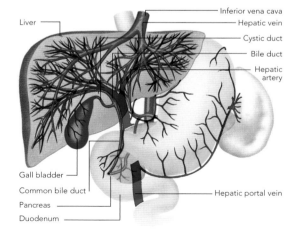

Liver

Inferior vena cava
Hepatic vein
Cystic duct
Bile duct
Hepatic artery

Gall bladder
Common bile duct
Pancreas
Duodenum
Hepatic portal vein

*blood leaves the liver through the hepatic veins that immediately empty*

*into the inferior vena cava vein which returns the blood to the heart.*

# KIDNEYS

The two kidneys play a vital role in homeostasis by processing the blood to produce a waste fluid called urine which is expelled from the body. By so doing, the kidneys are carrying out two major functions. The first is excretion of unwanted, and potentially poisonous, metabolic wastes produced by cells, notably urea, a nitrogenous waste product generated by the deamination of amino acids in the liver. The second is osmoregulation, the removal of excess water and salts (such as sodium and potassium salts) in order to maintain constant levels of water and salts in the blood and other body fluids.

## Kidney structure
The two bean-shaped kidneys lie high in the posterior abdominal cavity, partially protected by the lowest ribs. An adult kidney is about 5 inches (12 cm) long and 2.4 inches (6 cm) wide, and is protected by a thin outer capsule. On the medial side of each kidney is an indentation (or hilus) through which nerves and a renal artery enter, and a renal vein and urethra leave, the kidney. A section though the kidney reveals that it consists of three zones: an outer renal cortex, a middle renal medulla, and a central renal pelvis. A flattened, funnel-shaped tube collects urine and directs it into the ureter.

## Nephrons and urine production
Running between the cortex and medulla of each kidney are about 1 million microscopic nephrons. These are the 'filtering units' of the kidney that produce urine. Each nephron consists of a glomerulus and a renal tubule, comprising a Bowman's capsule, proximal and distal convoluted

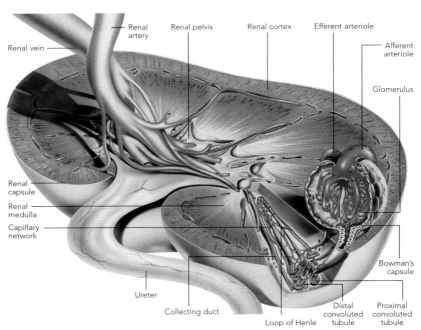

Renal vein — Renal artery — Renal pelvis — Renal cortex — Efferent arteriole — Afferent arteriole — Glomerulus — Renal capsule — Renal medulla — Capillary network — Bowman's capsule — Ureter — Collecting duct — Loop of Henle — Distal convoluted tubule — Proximal convoluted tubule

▲ **Kidney structure**
*This diagram shows a longitudinal section through a kidney, with a detailed,*

*magnified view of one the microscopic nephrons – each consisting of a glomerulus and a renal*

*tubule – found inside the cortex and medulla of each kidney.*

tubules, a loop of Henle, and a collecting duct (see diagram, below left). The glomerulus is a cluster of 'leaky' blood capillaries. Blood enters the glomerulus through an afferent arteriole and leaves through a narrower, efferent arteriole. The difference in diameters produces high blood pressure inside the glomerulus, which forces fluid (but not cells or plasma proteins) from the blood into the cup-shaped Bowman's' capsule. This fluid contains not just wastes, but also glucose, amino acids, water, and other material that the body needs to retain. As this filtrate passes along the proximal convo-luted (twisted) tubule, glucose and amino acids, and most salt and water is reab-sorbed into the blood flowing through the surrounding capillary network. Further concentration and reabsorption continues in the loop of Henle and the distal convo-luted tubule to produce a fluid, now called urine, that arrives in a collecting duct that receives input from several nephrons. The permeability of the col-lecting duct wall to water is mediated by the pituitary gland hormone ADH. In this way, urine can be further concentrated should the body need to retain more water, or left dilute if the body's water content is near to normal. Urine then passes into the renal pelvis and onward to the bladder for expansion (see pp. 96–97). About 95 percent of urine consists of water; the remaining 5 percent consists mainly of nitrogenous wastes, salts, and metabolized hormones. Its yellow color is produced by urochrome, a breakdown product of hemoglobin.

The importance of the kidneys in homeostasis is highlighted by the amount of blood they process daily. The renal arteries deliver about 25 percent of the cardiac output to the kidneys regardless of how active the body is; this amounts to about 2.5 pints (1.2 liters) per minute, or 3,600 pints (1,800 liters) per day. Each day 360 pints (180 liters) of fluid is filtered from blood plasma through glomerular capillaries and into Bowman's capsules; only one percent (about 3.5 pints/1.5 liters) leaves the body as urine. The body's entire blood supply is processed by the kidneys about 60 times every day.

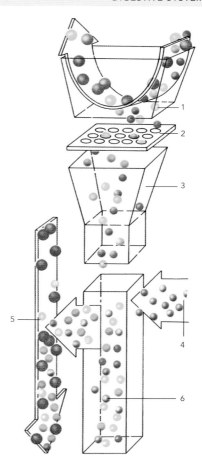

▲ The filtering mechanism
Blood flows at high pressure (above) through the capillaries of the Bowman's capsule (1) and only small molecules are forced through the walls (2) into the first part of the nephron (3). The filtrate passes down the proximal tubule, which secretes further metabolites and salts (4) and reabsorbs water, sodium, essential salts, glucose and amino acids into the blood (5). The loop of Henle and distal tubule are concerned with the reabsorption of water and maintenance of the overall acid-alkaline balance of the body. Unwanted salts, urea, and water are left as urine (6).

## Ureters, bladder, and urethra

Urine is produced by the two kidneys in a constant trickle of about 0.05 fl oz (1 ml) per minute. The other organs of the urinary system – the two ureters, the bladder, and the urethra – serve to channel and store urine until a convenient time is found for its release, and then transport urine to the outside of the body.

Each **ureter** is a hollow tube, between 10 and 12 inches (25–30 cm) long, with a wall that contains both longitudinal and circular layers of smooth muscle. It begins as an extension of the kidney pelvis and then travels vertically downward to open in the posterior wall of the bladder, an arrangement that prevents backflow of urine. Urine is pushed along the ureters by rhythmic contraction of the smooth muscle layers called peristalsis (*see* pages 84-85).

The **bladder** is a 'storage bag' that has a muscular, highly elastic wall. When empty, just after urination, it can be as small as a plum. As it fills, it can hold 10–13 fl oz (300–400 ml) of urine before a person feels a conscious desire to urinate, and can hold double that volume before the person has an urgent need to urinate. The wall of the bladder contains layers of smooth muscle fibers that contract to squeeze out urine.

The thin-walled, muscular **urethra** leaves the base of the bladder to carry urine to the outside. In adult males, it is about 8 inches (20 cm) long and opens at the tip of the penis. In adult females, it is about 2 inches (4 cm) long and opens halfway between the clitoris and the vaginal opening. In the bladder wall at its junction with the urethra is a ring of smooth muscle, the internal urethral sphincter, that is under involuntary control. Where the urethra passes through the pelvic floor muscles there is an external urethral sphincter that is made of skeletal muscle.

## Bladder filling and emptying

As the bladder fills with urine, its inner folds, or rugae, start to disappear. As this happens, stretch receptors in the muscular wall detect increasing tension in the wall and sends nerve impulses to the spinal cord. When the bladder contains 6–10 fl

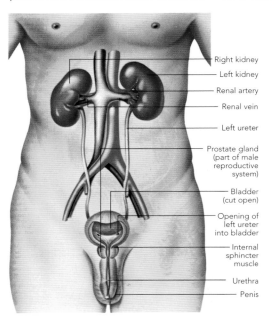

Right kidney
Left kidney
Renal artery
Renal vein
Left ureter
Prostate gland (part of male reproductive system)
Bladder (cut open)
Opening of left ureter into bladder
Internal sphincter muscle
Urethra
Penis

◀ *Urinary system*
*This diagram shows the organs that make up the male urinary system, a system that is located in the abdominal cavity. The two kidneys (kidney function is described on pages 94–95) lie in the upper, posterior abdominal cavity on either side of the backbone. Each receives oxygen-rich blood through a renal artery, a branch of the aorta, and returns oxygen-poor (but 'filtered') blood along the renal vein to the inferior vena cava. The muscular bladder lies in the lower, pelvic region of the abdominal cavity. Each kidney is connected to the bladder by a ureter which opens in the inferior, posterior surface of the bladder. The inferior exit from the bladder is guarded by a sphincter muscle, and leads to the urethra which carries urine to the outside.*

oz (200–300 ml) of urine, the spinal cord sends impulses to the bladder that stimulate the bladder wall to contract, and the internal urethral sphincter to relax so that urine passes into the top of the urethra. Shortly after this, messages traveling up the spinal cord to the brain give the person a conscious awareness that their bladder is filling. During urination, or micturition, the external sphincter muscle is relaxed under conscious control and contraction of the bladder wall pushes urine out through the urethra. This conscious control is learned during early childhood.

## Water content

Water is vitally important to the body because it provides the medium inside cells in which the reactions of life take place; and is a major component of extracellular fluids, such as tissue fluid and blood, that play a vital role in homeostasis by, for example, supplying oxygen, removing wastes, and keeping cells warm. For example, the body of a young man of average body mass (about 155 pounds/70 kg) will contain about 80 pints (40 liters) of water, with about 50 pints (25 liters) inside cells and 30 pints (15 liters) in the form of extracellular fluid.

The amount of water in a human body depends on a person's age, sex, and fat content. Adipose (fat) tissue contains less water than any other body tissue, including bone, so that the fatter a person is, the less water (in terms of percentage body mass) she or he will contain (see table). Infants, with low bone and body fat masses, are about 70 percent water. Young males contain less water than infants, but more water than young females because of the latter's higher fat content. Older people of both sexes contain less water because as people age their fat content increases.

| WATER CONTENT | |
|---|---|
| Age/sex | Water content (% body mass) |
| Infant | 72% |
| Young man | 59% |
| Young woman | 50% |
| Older man/woman | 45% |

Regardless of the age, sex, or fat content of the individual, it is essential to maintain stable water levels in both intracellular and extracellular 'compartments' of the body. Maintaining the water content at constant levels is one of the major roles of the kidneys.

## Water balance

Every day, an average adult takes in about 5 pints (2.5 liters) of water from both drinks and food as well as 'metabolic water' generated by aerobic cell respiration. Water is lost by sweating, in feces, directly across the skin and from the lungs, and in urine. While the table below shows the average amounts lost, the real proportions depend on how active a person is and the ambient temperature. Obviously, if someone sweats a lot, they will lose less water in their urine. This is mediated by the pituitary hormone ADH, the secretion of which increases if body fluids become more concentrated, thereby conserving water inside the body. This monitoring of the concentration of body fluids, particularly blood, is carried out by the hypothalamus. As well as causing the pituitary gland to release ADH should blood concentration increase, the hypothalamus also generates the conscious feeling of thirst, so that a person drinks water to dilute blood plasma. By varying the volume and concentration of urine released by the body, the kidney ensures that daily water intake is balanced by daily water output.

| TABLE: WATER BALANCE | | | | |
|---|---|---|---|---|
| Water intake | | Water output | | |
| Drinks | 3.2 pints/1,500 ml | Urine | 3.2 pints/1,500 ml |
| Food | 26 fl oz/750 ml | Lost through skin and lungs* | 24.5 fl oz/700 ml |
| Metabolic water | 8.75 fl oz/250 ml | Sweat | 7 fl oz/200 ml |
| | | Feces | 3.5 fl oz/100 ml |
| Total | 5.3 pints/2,500 ml | Total | 5.3 pints/2,500 ml |

* water vapor that passes out in exhaled air, and water that is lost directly across the skin.

# MALE REPRODUCTIVE SYSTEM

Much of the male reproductive anatomy is external. The two testes hang in the scrotal sacs surrounded by the tunica vaginalis. The vasa efferentia join the testes to the overlying epididymis and vas deferens, which join the urethra in the center of the prostate gland. Akin to the gall bladder, the seminal vesicle acts as a storage organ for the mature sperm. The seminal vesicle lies between the prostate gland and the colon.

The wrinkled subcutaneous area of the scrotum contains the dartos muscle, which can contract and, with the cremaster muscle, attached to the spermatic cord, pull the testes closer to the abdomen. The spermatic cord is the combination of the vas deferens (surrounding testicular blood vessels and nerves), fatty tissue and the projection from the peritoneum that originally formed the tunica vaginalis.

## The prostate gland

The prostate gland lies around the first part of the urethra at the base of the bladder, and its secretions help maintain sperm activity. The paired seminal vesicles empty into the urethra adjacent to the prostate and contain energy-rich sugars that make up 60 percent of the 0.15 to 0.25 fl oz (3–5 ml) in each ejaculate. The paired bulbourethral glands at the base of the penis provide an alkaline liquid that makes the urethra less acidic and more 'sperm-friendly.'

Semen contains various proteins, fructose, and a mixture of chemicals, which help with the nutrition of sperm in the vagina. Each 0.3 fluid ounce contains about 100 million sperm, of which 20 percent are dead or abnormal. The number of sperm produced at ejaculation is variable.

The penis surrounds the fibroelastic urethra, which runs in the corpus spongiosum and ends in the glans, which is covered by the foreskin. The urethra and corpus spongiosum are covered, above and at the sides, by the pair of

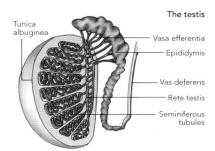

The testis

corpora cavernosa. Normally the penis is limp, but if the muscles at the base contract around the veins, the spongy tissue in all three corpora becomes congested with blood and an erection occurs. The penis has two functions: the excretion of urine from the bladder out of the body; and the deposition of semen in the vagina of the female genital tract.

## The testis

The testis has two functions: the production of testosterone and the production of spermatozoa. It is about 2 inches (5 cm) long and 1 inch (2.5 cm) thick and surrounded by the tunica albuginea; this coat is divided into about 200 lobes containing 4 to 600 seminiferous tubules, each about 30 inches (75 cm) long. The

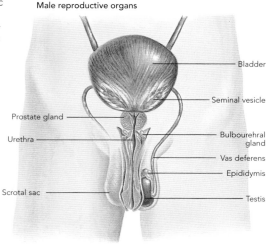

Male reproductive organs

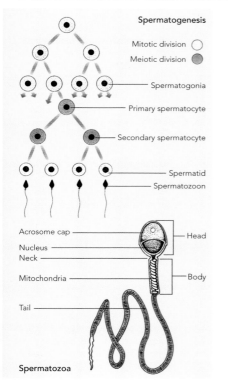

**Spermatogenesis**

Mitotic division ○
Meiotic division ◉

Spermatogonia
Primary spermatocyte
Secondary spermatocyte
Spermatid
Spermatozoon

Acrosome cap ——— Head
Nucleus
Neck
Mitochondria ——— Body
Tail

**Spermatozoa**

tubules produce about 200 million sperm every day which pass into a series of communicating ducts – the rete testis. These ducts, lined with ciliated cells, join to form 12 to 20 vasa efferentia, which run to the epididymis, where the sperm spend up to ten days maturing before entering the vas deferens. The sperm can then survive a further six weeks before degeneration and absorption.

Spermatogenesis requires stimulus from the follicular stimulating hormone of the anterior pituitary gland. Testosterone responds to luteinizing hormone. It causes development of male secondary sexual characteristics – pubic and facial hair growth, aggressiveness, muscle bulk, and deepening of the voice.

### Spermatogenesis

The spermatogonia divides by **mitosis** to form the primary spermatocyte, containing 46 chromosomes. **Meiosis** occurs when half of each chromosome pair enters the secondary spermatocyte. This divides again in order to produce a spermatid that finally matures into a spermatozoon.

### The spermatozoa

The mature sperm is 0.0020 inches (0.05 millimeters) long. It consists of a head, body, and tail. The head is covered by the acrosome cap and contains a nucleus of dense genetic material from the 23 chromosomes. It is attached by a neck to a body containing mitochondria that supply the energy for the sperm's activity. The tail is made of protein fibers that contract on alternate sides, giving a characteristic wavelike movement that drives the sperm through the seminal fluid, which also supplies additional energy. Some sperm have two heads or two tails and, if the testes are too warm, they may die or spermatogenesis may not occur.

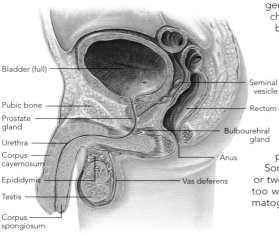

Bladder (full)
Pubic bone
Prostate gland
Urethra
Corpus cavernosum
Epididymis
Testis
Corpus spongiosum
Seminal vesicle
Rectum
Bulbourehral gland
Anus
Vas deferens

# FEMALE REPRODUCTIVE SYSTEM

The female reproductive system not only has to produce an ovum but has then to nurture the fertilized ovum and protect it until pregnancy ends. At the entrance to the vagina, a pair of liplike folds – the larger, thicker labia majora and the smaller, inner labia minora – lie along either side of the vaginal entrance and join in the front, blending into the padded, hairy area of the mons pubis. At the front, they enclose the exit of the urethra just behind the small projection of erectile tissue – the clitoris – comparable to the male penis.

Behind these structures is the vagina – a 4 to 6 inch (10 to 15 cm) -long elastic tube lined with moist epithelium; it is normally 'closed' except during intercourse. At the back of the vulva are two large Bartholin's glands, and lining the entrance to the vagina are a large number of smaller, lubricating glands. At the top of the vagina, the uterus is held in place by the muscles and four strong fibrous ligaments of the pelvic floor, and to the side of the pelvis by pairs of the round and suspensory ligaments, running in folds of peritoneum.

## The uterus

The uterus is a small pear-shaped organ, covered with peritoneum, with a thick wall of interweaving muscle fibers and lined with the special endometrial cells. It is located behind the bladder and in front of the rectum. The cervix of the uterus is a thick fibromuscular structure opening into the vagina and lined with special cells that form a 'plug' of mucus. The uterine muscle is always contracting and relaxing slightly. This wavelike movement increases during orgasm in order to 'suck in' sperm; during menstruation in order to expel the endometrium; and also at parturition.

The two Fallopian tubes are about 4 inches (10 cm) long with fingerlike fimbriae at the ends to encircle the ovaries. A combination of ciliated epithelium and peristaltic muscular contractions

### The ovary

*This cross section of the ovary shows an immature ovum (1) absorbing fluid and swelling into the saclike Graafian follicle (2). As it reaches the surface (3), half way through the menstrual cycle, the ovum bursts out into the peritoneal cavity to be collected by the fimbriae.*

*The empty follicular cyst then changes into the corpus luteum (4), which grows until menstruation, when it shrivels into a scablike corpus albicans (5).*

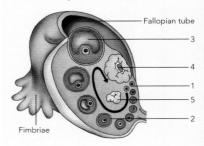

sweeps the ovum down the Fallopian tube. Unlike the male urinary system, that of the female is separate from the reproductive system. The bladder empties into the urethra, which opens in front of the vagina.

Female Reproductive organs

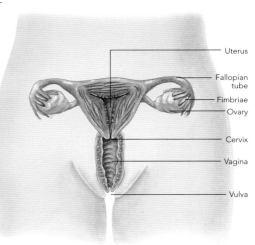

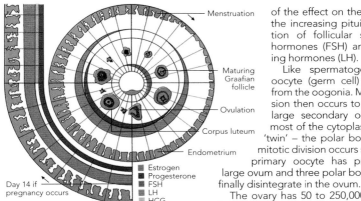

Menstruation

Maturing Graafian follicle

Ovulation

Corpus luteum

Endometrium

Day 14 if pregnancy occurs

■ Estrogen
■ Progesterone
■ FSH
■ LH
■ HCG

The menstrual cycle

of the effect on the ovaries of the increasing pituitary secretion of follicular stimulating hormones (FSH) and luteinizing hormones (LH).

Like spermatogenesis, an oocyte (germ cell) is formed from the oogonia. Meiotic division then occurs to produce a large secondary oocyte with most of the cytoplasm from its 'twin' – the polar body. Further mitotic division occurs so that one primary oocyte has produced a large ovum and three polar bodies, which finally disintegrate in the ovum.

The ovary has 50 to 250,000 oogonia, but only about 500 eventually become mature ova.

## The ovary

The ovary has two functions: the production of ova (see diagram, top left); and the secretion of estrogen and progesterone. The ovary is about 1 inch (2 cm) across and 0.4 inch (1 cm) thick.

At puberty, the onset of menstruation and the development of the secondary sexual characteristics – hair growth, breast development and redistribution of fat to the buttocks and shoulders – are the result

## The menstrual cycle

From menarche to the menopause, the anterior pituitary lobe maintains a rhythm of FSH and LH secretion to produce, in most women, a regular menstrual cycle (see diagram, right). FSH stimulates the maturation of several Graafian follicles, of which only one reaches maturity. The follicular cells produce estrogen to build up the endometrium. In midcycle, a surge of LH softens the plug of cervical mucus and causes ovulation. The corpus luteum is formed and secretes progesterone, which, with estrogen, further prepares the endometrium for implantation of the fertilized egg.

If fertilization takes place, the embryo produces human chorionic gonadotrophin to maintain the stimulus to the corpus luteum and the continued production of progesterone. If fertilization does not occur the corpus luteum degenerates, progesterone production ceases and the endometrium is shed, causing menstruation for about five days. The change in the hormonal balance may be responsible for premenstrual syndrome.

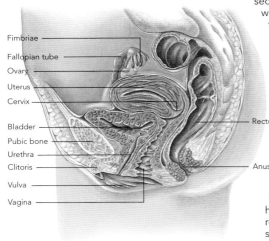

Fimbriae

Fallopian tube

Ovary

Uterus

Cervix

Bladder

Pubic bone

Urethra

Clitoris

Vulva

Vagina

Rectum

Anus

# FERTILIZATION AND IMPLANTATION

## Coitus

Coitus is the insertion of the erect penis into the vagina, followed by rhythmical movements ending in orgasm and the ejaculation of semen by the man. Initially sexual excitement is produced by sight, touch, sound and, perhaps, pheromones that stimulate smell. In the man, there is congestion of the corpora cavernosa and corpus spongiosum of the penis. In the woman, there is slight engorgement of the breasts, and congestion of the clitoris and labia, with increased vaginal secretions. A prolonged phase of enjoyment can then follow when the penis is inserted into the vagina, provided the 'crescendo' to orgasm is controlled.

Orgasm is a series of rapid muscular contractions surrounding the male urethra that result in the ejaculation of the semen. This can be complemented by a simultaneous experience in the woman, in which the upper part of the vagina and a rhythmical 'sucking in' of the uterus draw the sperm to the right area. During sexual intercourse there are increases in blood pressure, rates of heartbeat, and respiration. After orgasm there is physical relaxation.

## Fertilization

The moment of conception is the most important stage of sexual reproduction. The joining of the two nuclei, each containing 23 chromosomes, to form a cell of 46 chromosomes, leads to the creation of an embryo.

The moment of conception

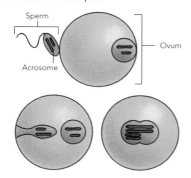

The sex of the baby is decided by the father's chromosomes – the XY sex pair. When the primary spermatocyte divides by meiosis to form the secondary spermatocytes, containing 23 chromosomes, either an X or a Y chromosome will move into each cell. The female pair of chromosomes are the same, XX, so all ova will contain the X chromosome. The fertilizing sperm will join the ova to form an XY male or XX female embryo. It is probable that there is less than 24 hours in which the ovum can be fertilized and as the sperm survive for about 48 hours in the uterus and tubes, there is only a limited time each month during which conception is able to take place.

## The sperm's journey to the ovum

The ejaculate of semen contains more than 350 million sperm. Ideally they are deposited adjacent to the cervix, where the vaginal enzymes dissolve the seminal mucus and release the sperm. They then swim into the cervical mucus, which is watery and soft enough at ovulation to allow many of the sperm to penetrate to the uterine cavity. Uterine contractions will then help them to move upward to the Fallopian tubes, where they again swim, perhaps with the help of the lining of ciliated cells, to reach the ovum at the midpoint of the tube. Only about 100 sperm survive the journey of nearly 24 hours and only one fertilizes the ovum. The ultimate fate of the surplus spermatozoa in the female genital tract is still not known.

The sperm's acrosome disappears as it dissolves the membrane of the ovum. The tail and body are shed when the head penetrates to join its 23 chromosomes with those of the ovarian nucleus.

## Implantation of the ovum

It takes about a week for the fertilized ovum to pass down the Fallopian tube and implant itself in the endometrium. The following description of implantation complements the diagram on the right.

Within hours of conception, mitosis begins with the development of a sphere of an increasing number of cells. The sphere starts as the blastomere (1), then becomes a morula (2) of about 64 cells. At this stage it changes into a hollow, fluid-containing ball – blastocyst (3) – with the inner cell mass at one end. It can now begin implantation (4) into the lining of the uterus. The outer ring of cells, now called the trophoblast, secretes enzymes that erode the endometrium. The trophoblastic cells spread into the endometrium, forming lacunae (fluid-bearing sacs) that penetrate the maternal circulation. This allows nutrition to take place. Ultimately, the trophoblast forms the outer layer of the placenta.

The inner cell mass splits into two layers: endoderm – ultimately the alimentary tract – which produces a yolk sac; and the ectoderm – ultimately the skin, brain and spinal cord – which produces the amniotic sac to surround and protect the embryo in a bag of fluid.

By the ninth day after conception, the blastocyst has sunk deep into the endometrium (5) and is already receiving nutrition from the mother. The trophoblast secretes human chorionic gonadotrophin in order to maintain the progesterone production from the corpus luteum. This will continue until the end of the third month, when the placenta produces sufficient hormone to maintain this function until the end of pregnancy.

Implantation of the ovum

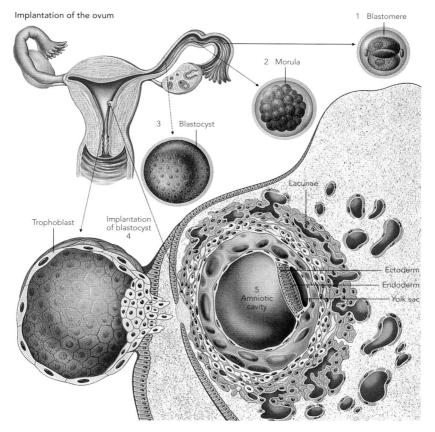

# PREGNANCY

Growth of embryo

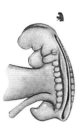 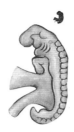 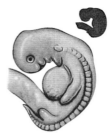 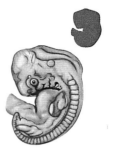

Three-week embryo

Four-week embryo

Five-week embryo

Six-week embryo

## The first four weeks

The blastocyst begins to implant into the endometrium on the seventh day after conception. It consists of two distinct parts: a covering layer of the trophoblastic cells and an inner cell mass.

The trophoblast invades the maternal tissues and erodes the maternal blood vessels so that the surface is bathed in oxygen-carrying nutritious fluid. In addition the trophoblastic cells produce human chorionic gonadotrophin to stimulate the production of progesterone and to help maintain pregnancy.

The inner cell mass develops a fluid-containing cavity from its upper surface (the amniotic sac), which eventually expands to line the inner surface of the trophoblastic cell layer. This layer ultimately forms the chorion of the fetal membranes, and protects the embryo and fetus until parturition (childbirth).

## The next two months

By the end of the third week (see artwork above), the mesodermal tissue begins to arrange itself into somites – a segmental arrangement that can be seen in humans in the ribs of the thoracic cage. Each somite of muscle is eventually supplied by its own blood vessels and nerves.

At four weeks, the embryo is 0.125 inches (3 millimeters) long. It consists of 25 somites, a large bulge of the heart, and has small ear pits in the head. It is not until the fifth week that a pair of pig-mented disks appear on the head. These are the first signs of the eyes. In front is the small depression of the primitive nose. The limb buds, arms first and then legs, begin to grow by the sixth week.

At six weeks, the embryo has a tiny pulsating heart, forty somites and miniature ear flaps. At this stage, the embryo is nearly 0.5 inches (1.5 centimeters) long and the hands, fingers, and feet are just about discernable.

Next, the somites lose their well-defined divisions and join up to form a solid mass of body tissue with recognizable thorax, abdomen, and the beginnings of a neck. By the seventh week, the bulbous protuberance is clearly a head with well-defined nose, ears, eyes, and mouth. The pelvis is starting to form and the end somites lose their taillike appearance as the sacral bone is developed.

It is during this rapidly changing period, from the fourth to the seventh week, that congenital malformations may occur if the tissues fail to fold or divide correctly. External influences, such as disease or drugs, may damage the embryo.

By the eighth week, the embryo is approximately 1 inch (2.5 centimeters) long and is recognizably human. It has distinct eyelids and separate fingers and its toes are developing.

From the end of the second month onward, development is principally by growth and some alteration in proportion. By the end of 12 weeks, the embryo is 2

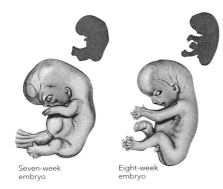

Seven-week embryo

Eight-week embryo

is 5 inches (12.5 centimeters) long. The genitalia are developing and ossification centers appear. The fetus is moving in the amniotic fluid, and in the fifth month (see below left) the mother may feel these movements – known as 'quickening' – for the first time.

In the sixth month, the fetus starts to produce a layer of fat under the skin and a fine hair – lanugo – covers the head and body, which is also covered by a greasy substance – vernix – that protects and lubricates the fetal skin, particularly during parturition. As the fetus grows, the proportion of amniotic fluid decreases. At 28 weeks, the fetus weighs about 17.5 ounces (500 grams) and its heart-beat of 140 a minute can easily be heard; the fetus is now viable. Four to six weeks later, the head engages in the pelvis and remains there until parturition. During the last two months (see below right), it matures steadily and gains about 1 ounce (25 grams) weight per day to reach about 6.5 pounds (3 kilograms) at birth.

inches (5 centimeters) long and the mother's uterus is about the size of an orange. It is at this stage of development that the embryo is called a fetus.

## The continuation of pregnancy
At the end of the fourth month, the fetus weighs about 4.5 ounces (125 grams) and

Five-month fetus

Nine-month fetus

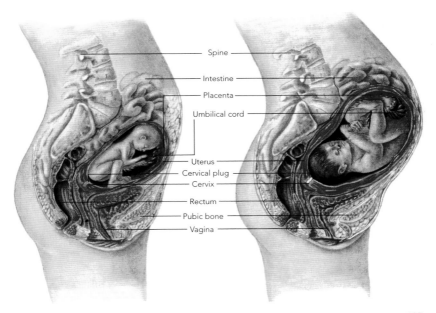

Spine
Intestine
Placenta
Umbilical cord
Uterus
Cervical plug
Cervix
Rectum
Pubic bone
Vagina

# CHILDBIRTH

At about 280 days after conception, the fetus is expelled from the uterus. If pregnancy continues for too long the placenta begins to degenerate and the fetus is slowly starved of oxygen and nutrition.

Throughout pregnancy the uterine muscle has been contracting and relaxing. During the last few weeks, these contractions become stronger and are sometimes noticed by the mother. They help to keep the fetal head engaged in the pelvis (A). Labor is divided into three stages: the first is from the onset of labor, with regular rhythmical uterine contractions, to the moment when the cervix is fully dilated. This first stage may last from two to 24 hours or sometimes longer. The second stage is from the full dilation of the cervix to the moment of delivery and lasts less than two hours. The third stage is the expulsion of the placenta, cord, and membranes. The placenta has to be carefully examined to make sure that parts of it have not been left behind which could cause a hemorrhage. The stages are described in greater detail below.

## The first stage

The first stage starts with backache and regular uterine contractions that become gradually stronger and by the end occur about every two minutes, each lasting 30 to 40 seconds. As the head is forced deeper into the pelvis, it turns sideways (B) and the plug of cervical mucus is dislodged causing a 'show' of blood. The cervix is gradually stretched, thinned and dilated until it is wide enough (C) for the head to pass into the vagina. If it has not already done so, the amniotic sac ruptures.

## The second and third stages

The second stage starts with the cervix fully dilated (C). The head moves deeper into the pelvis and turns again, so that the occiput is behind the symphysis pubis. The mother has a 'bearing-down' sensation and can assist the powerful uterine contractions to force the fetus through the vagina. The vulva distend around the head and finally the baby is 'crowned' when the greatest circumference of the vulva is reached. The head extends (D) around the symphysis as it is

Sequence of normal birth

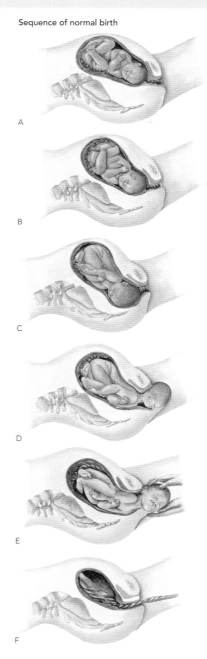

A

B

C

D

E

F

pushed out. The baby takes its first breath and lets out a cry. This expands the lungs and starts respiration, as well as obliterating the fetal circulation.

The baby can now be delivered (**E**) with the next contraction as the smaller shoulders and body easily follow the large head. At this moment the mother is usually given an injection of oxytocin and ergometrine to stimulate uterine contraction and the separation of the placenta.

The third stage follows ten to twenty minutes later when the placenta has separated from the uterus and is expelled (**F**), along with the umbilical cord and membranes.

## The puerperium
After parturition the uterus returns to normal size in about a month with loss of lochia, which is initially fresh blood later becoming creamy in color. Menstruation will return in about two months unless suppressed by prolactin secretion during lactation. Excess body fluid and fat gradually disappear and softened ligaments regain their strength.

The baby is born with excess fluid, which it loses in the first few days, and a raised hemoglobin count level, which falls slowly with the appearance of slight jaundice. This is the result of raised levels of serum bilirubin from hemolysis and the immature liver's inability to excrete it.

## Lactation
Throughout pregnancy the breasts have been enlarging under the stimulus of the anterior pituitary hormone, prolactin, to produce a watery milk (*see* diagram, right). In the last few weeks before parturition colostrum is produced and this is secreted for the first few days after birth. After this brief period, the mother's breasts begin to produce normal milk (**1**) and the baby's sucking (**2**) stimulates the hypothalamus. This continues the production of prolactin and also oxytocin from the posterior pituitary gland (**3**), which causes the breast alveoli to contract and force – 'let down' – the milk into the ducts (**4**).

Hormones in labor

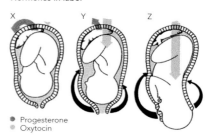

● Progesterone
○ Oxytocin

## Hormones in labor
The story of hormones in labor (*see* diagram above) is not fully understood. In some way their interreaction adjusts the actual onset of labor. Throughout the term of pregnancy, the production of progesterone makes the uterine muscle (**X**) relaxed and unresponsive to the posterior pituitary hormone – oxytocin. Toward the end of pregnancy, production of progesterone falls and there seems to be an increase in estrogen. The combination of oxytocin and estrogen initiates uterine contractions. The dilatation of the cervix (**Y**) increases oxytocin output and further reduces progesterone, so the contractions become stronger, longer and more frequent (**Z**). In the third stage, oxytocin helps to obliterate the spiral arteries in the uterine wall and initiate lactation. Oxytocin can be synthesized and is often used to start contraction of the uterus if labor is delayed.

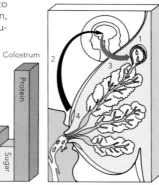
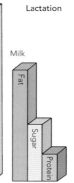

Lactation

Colostrum

Milk

## Infancy

A newborn baby lies with its knees drawn up. It reflexly grasps any object that touches the palm and, when held upright, automatically steps as the feet touch something. It roots and sucks the nipple automatically. These reflexes disappear within a few weeks. At one month, the legs are straighter and by six weeks the head is lifted. The baby sleeps more often than not, but gradually the eyes move to focus on objects and, at about six weeks, smiling begins. By six months, the child's weight has doubled since birth and it sits unaided. At eight months, the preliminary gurglings of speech are heard and the thumb can be used. At about ten months, crawling starts. The child's weight has trebled since birth. The first step may be taken at age one. The first words may be spoken during the next two or three months.

## Teeth

Deciduous teeth start to appear at about six months. By age three, the child has 20 teeth (see diagram below): eight incisors, four canines and eight premolars.

Between the ages of six and 12 years, the deciduous teeth are shed and replaced by the permanent dentition (see diagram, page 85). A further six teeth, the molars, will appear in each jaw, and by the age of 25, a total of 32 teeth will be present.

The tooth consists of three parts – the crown, neck, and root. The **crown** consists of dense mineral (enamel) surrounding the hard dentine, which has a soft center – the pulp. The pulp is filled with blood vessels, lymphatics, and the nerve, which reach it through the root canal. The **neck** of a tooth adheres to the gum, while the **root** penetrates the bone, where it is held in place by a ligament and cementum.

▲ *By the age of three, a child has 20 'milk' teeth – eight incisors, four canines, and eight premolars.*

## Puberty

Puberty is the moment when sexual maturation has reached the point where sexual reproduction is possible. A period of development around puberty takes about two years and will usually start earlier in girls than it does in boys.

In girls, between the ages of 10 and 16, there are changes in the subcutaneous fat – hips and shoulders become more rounded, the breasts develop and the pubic and axillary hair grows. The first menstruation – menarche – establishes a definite point of maturity.

In boys, between the ages of 12 and 17, the shoulders broaden, the muscles strengthen and the genitalia develop and darken to become covered with pubic hair. The larynx lengthens and the voice 'breaks' to become deeper. Spontaneous erections occur and the nocturnal emissions of sperm are perhaps less dramatic a sign of sexual maturity than the menarche.

## Adolescence

Adolescence is a time of great physical and emotional change, and can cause anxiety without the appropriate care and guidance. It is a time when social, intellectual, and sexual interests expand and broaden. Adolescents may experiment with different life styles and identities. The young person is no longer a child. She or he is more or less physiologically mature, but may feel frustrated by the constraints on their freedom in comparison with adults.

The intense emotions of adolescence can be expressed by the individual in a number of different ways, mostly harmless. However, adolescents often experience dramatic mood swings. Periods of great enthusiasm may be followed by perioids of almost depressive lethargy.

## The young adult

As the turmoil of adolescence settles, young adults (18–26) are believed to be at their physical prime. In addition to body maturity, relationships with family and society may be more positive and there is generally a greater awareness and control of emotions and sexual desire. The first tentative adolescent meetings with the opposite sex have passed and mature relationships develop. Entrance into the world of work may provide an element of independence and many young adults may find greater self-confidence.

## Middle age

The most stable period of life is usually considered to start around the age of 30. It can be a time of relative financial and emotional security with greater emphasis placed on relationships. Sexual desire is less urgent but often more satisfactory and women are considered to be at their sexual peak.

In women, the menopause – the final cessation of menstruation – generally starts around the age of 50. It may be followed by a time of hormonal disturbance leading to hot flushes, tensions, and sometimes depression. A program of hormone replacement treatment (HRT) can help to ease many of the problems that accompany the menopause.

Although physical health usually remains good, there is a tendency to gain weight, accompanied by a slight, but definite loss of athletic skills. Men may begin to experience hair loss. In the later years there is an increase in the death rate, particularly from heart disease and cancer of the lungs.

## Old age and senescence

The definition of 'old age' varies from society to society and also between individuals. In a general manner, however, it can be characterized by changes in close familial relationships, financial status and personal health. Common health issues include arthritis, and a gradual loss of visual acuity and hearing, which may lead to a slower reaction time.

As we age, the adrenal glands continue to secrete sex hormones and the slight androgenic effect may produce slight hair loss in women, while in men there may be the estrogen effect of female fat distribution and less beard growth. Cancer and other illnesses are common and often fatal.

► *Four female generations of a family, from the great-grandmother at far right to great-granddaughter at far left.*

Senescence is the period of real aging and may begin long before or long after the arbitrary point of 'old age.' The skin loses its elasticity and gains wrinkles. The hair loses its color, becoming white as it fills with microscopic bubbles of air. The senses of vision, hearing, and taste deteriorate. The joints become rough and arthritic, while the muscles weaken. The bones in the skeleton soften as the structure deteriorates and calcium is removed, causing the bent spine of many of the elderly.

The nervous system begins to lose nerve cells, leading to failure of memory – initially for recent events but later a gradual dementia may occur. The cerebellum is affected, producing a loss of coordination, tremor of the hands, and hesitation in speech. The heat center in the thalamus is less responsive and a cold environment may cause a drop in body temperature, a condition known as hypothermia, that can lead to coma and death. Deterioration in the heart and lungs may cause irregularities of cardiac rhythm and often breathlessness.

In some people these aging processes may take place faster in one system than another. This can be seen with dementia occurring while the body is fit, or osteoarthritis crippling an otherwise perfectly healthy person.

In some cases, the pressures of senescence are enormous and the individual may need professional help. In extreme cases, admission to hospital is the only way of giving sufficient basic nursing care.

# GENES AND INHERITANCE

## Heredity

In normal circumstances, a cell divides and the two new cells contain the same number of identical chromosomes. This is known as **mitotic division**. Human cells contain 23 pairs of chromosomes.

Reproductive cells, known as gametes, form by the splitting of the chromosome pairs to produce 23 chromatids in each cell. This is known as **meiotic division**. If this were the only process in operation, all the children of a particular couple would be identical.

Before meiosis occurs an apparently normal mitotic cell division takes place, but the chromosomes overlap and seem to 'stick' to each other. When they are pulled apart by the cytoplasmic network, the chromosomes break and change places with each other, thus altering the genes. This is known as 'crossing over' (see page 111).

As each gamete or reproductive cell is different, this alteration of genes gives each person a unique individuality. There are many features that will be nearly identical in close relatives and this makes members of the family recognizable as relations. Identical twins occur when the zygote – fertilized ovum – splits mitotically, producing two identical cells.

When the chromatids meet to form new pairs, known as alleles, certain physical features are opposite each other and, like the chromosome, become paired. There are probably 50,000 pairs of genes in each zygote carrying all the future instructions for the body. Many of these pairs will carry identical instructions, known as homozygous, but if they are different or heterozygous, one gene will have 'stronger' instructions than the other. This is known as the dominant

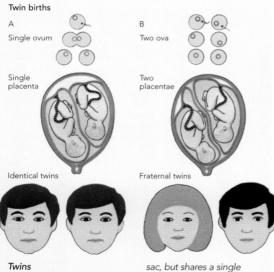

### Twins

Twin births occur about once in every 85 pregnancies. Identical twins (A) are always the same sex, developing from mitotic division of the same fertilized ovum. Each fetus has its own amniotic sac, but shares a single placenta. Fraternal, nonidentical twins (B), develop from two separate ova that have been fertilized at the same time by different spermatozoa. They have separate placentae.

▲ *The sex of an individual*
*An individual's sex is determined by a particular pair of chromosomes. If they are homozygous, a female is produced and the two alleles are of equal length with the same number of genes. These are called XX. If the pair is heterozygous, XY, a male is produced. The Y allele is smaller and contains fewer genes than the X half of the pair. Thus the unpaired genes act in a dominant manner.*

gene. The weaker one is known as the recessive gene. Some genes are not always dominant or recessive; one may only tend to dominate.

The majority of genes do not vary and are homozygous. Most people have the same number of features and grow in a very similar manner. If the genes are badly damaged, it is unlikely that the zygote will develop at all. Minor genetic changes may produce abnormalities, usually harmful but occasionally beneficial.

## Gamete formation

Gamete cells are produced to transmit genetic information (see diagram, right) from parents to their children.

Chromosomes appear in the nucleus (1). The DNA splits and the chromosomes are duplicated as the nuclear envelope breaks (2).
Similar chromosomes overlap and adhere to each other – 'crossing over' – and exchange sections (3).
The cytoplasmic network pulls apart the chromosomes and lines them up in the center (4).
The chromosomes are then pulled to opposite poles of the cell (5).
The cell membrane breaks and two new cells are formed, each containing 46 nonidentical chromosomes (6).
Meiosis can now start (7).

The two chromosome pairs split apart and each half, a chromatid, is pulled by the cytoplasmic network to the side of the cell (8).
The cell membrane breaks and the two new pairs of cells are formed, each containing 23 chromatids. This is a gamete.

The joining of two gametes causes pairing of the chromatids and formation of a new cell, the zygote, with 46 chromosomes (9). Occasionally, more or less than 46 chromosomes are formed, which usually causes a severe disorder. For instance, the condition known as Down's syndrome is caused by the presence of an extra copy of chromosome 21.

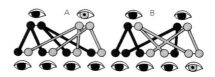

▲ Eye color
There is more than one gene for eye color, but brown is dominant over blue. Two people, one with two genes for brown eyes, the other with two genes for blue eyes (A), will have children who all have brown eyes. However, if two brown-eyed parents (B) carry the heterozygous recessive blue gene, they will have one blue-eyed child for every three brown-eyed children.

Meiosis – gamete formation

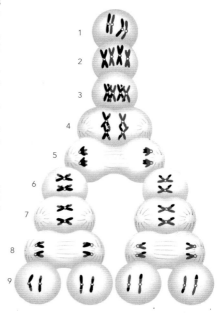

Chromosomes crossing over

**A**

Supreme Court established a woman's right to have an abortion during the first three months of pregnancy, except under certain defined conditions. This affected the abortion laws in 31 States.

**abscess** Collection of pus anywhere in the body, contained in a cavity of inflamed tissue. It is caused by bacterial infection.

**acetaminophen** ANALGESIC drug that lessens pain and is also effective in reducing fever. It is used to treat mild to moderate pain, such as headaches, toothaches, and rheumatic conditions, and is particularly effective against musculoskeletal pain.

**acidosis** Abnormal condition in which the acidity of the body tissues and fluids is unduly high. It may arise in a number of conditions, including kidney failure, severe diabetes, shock, and some forms of poisoning. Symptoms include breathlessness, weakness, and general malaise.

**acne** Inflammatory disorder of the sebaceous (oil-producing) glands of the skin resulting in skin eruptions such as blackheads and infected spots; it is seen mostly on the face, neck, and back. Acne is common in both sexes at puberty. It does not

**abortion** Termination of pregnancy before a fetus is sufficiently advanced to survive outside the mother's uterus. **Spontaneous** abortion, better known as a miscarriage, occurs in c.20% of apparently normal pregnancies. Miscarriages in the first three months of pregnancy are usually caused by fetal abnormalities. Miscarriages later in pregnancy may be caused by defects in the maternal environment, such as reproductive system disorders. Induced or therapeutic abortion is the termination of pregnancy by drugs or surgery. The rights of the fetus and the mother's right to choose provoke much political and ethical debate. In *Roe v. Wade* (1973), the US

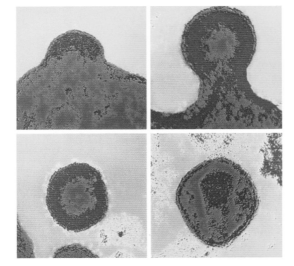

◄ *AIDS The sequence of false-color transmission electron micrographs shows the formation of an HIV particle at the surface of an infected lymphocyte. The first appearance of the virus (top left) is as a small bump (red) on the surface of the cell. The virus then buds out (top right) and is eventually cut off from the cell membrane. The newly released virus particle (bottom left) has an outer shell of dense material (red) which is reorganized to form the central nucleoid that becomes characteristically elongated in the mature virus (bottom right). Magnification: ×100,000.*

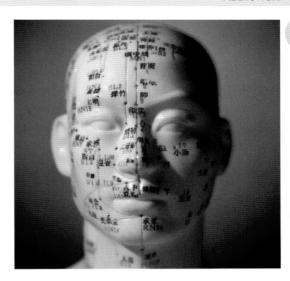

▶ *Acupuncture* Man's head marked with a acupuncture chart. Meridians (lines) and points are mapped out across the skin. The meridians relate to channels of energy within the body. Imbalances in the flow of energy are thought to cause pain and disease. The insertion of needles at precise points (marked by dots), at different angles, temperatures, and speeds according to the disorder, is thought to stimulate energy flow. Acupuncture may work by inducing the production of natural analgesics within the central nervous system, or by stimulating nerves which inhibit the original pain.

**A**

usually persist beyond early adulthood. Severe cases can be treated with drugs.

**acquired immune deficiency syndrome (AIDS)** Fatal disease caused by a retrovirus, called HUMAN IMMUNODEFICIENCY VIRUS (HIV), that mainly attacks T-4 cells (which help the production of antibodies) and renders the body's immune system incapable of resisting infection. The first diagnosis was made in New York in 1979. In 1983, scientists at the Pasteur Institute in France and the National Cancer Institute in the USA isolated HIV as the cause of the disease. The virus can remain dormant in infected cells for up to 10 years. Initial **AIDS-related complex (ARC)** symptoms include severe weight loss and fatigue. It may develop into the AIDS syndrome, characterized by secondary infections, neurological damage, and cancers. AIDS is transmitted only by a direct exchange of body fluids. Transmission is most commonly through sexual intercourse, the sharing of contaminated needles by intravenous drug users, and the uterus of infected mothers to their babies. In the United States and Europe, more than 90% of victims have been homosexual or bisexual men. However, 90% of reported cases are in the developing world, and many victims are heterosexual. AIDS is now the leading cause of death in sub-Saharan Africa. Recent combinations of drugs have met with some success in controlling symptoms. By 2002, c.25 million people had died from AIDS.

**acupuncture** System of medical treatment in which long needles are inserted into the body to assist healing, relieve pain, or for anesthetic purposes. In ancient Chinese philosophy, acupuncture is proposed to restore the balance of yin and yang by freeing the flow of life-energy (chi) through pathways in the body. A possible scientific explanation is that the needles activate deep sensory nerves that stimulate the pituitary gland and hypothalamus to produce endorphins.

**addiction** Inability to control the use of a particular substance, resulting in physiological or psychological dependence. It is most frequently associated with drug addiction. In a medical context, addiction requires a physical dependence. When the dose of a drug is reduced or withdrawn, the addict may experience withdrawal syndromes. Psychological dependence on activities such as gambling or exercise are difficult to distinguish from a disorder or mania.

113

**A**

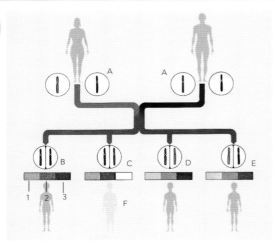

◄ *Albino A male and female both carrying a recessive gene (green) for albinism (A) will have three normally pigmented children (B, D, and E) to every one albino (C). The corresponding normal gene (orange or purple) in both produces normal skin color. This is due to an amino acid, phenylalanine (1), which has been converted to tyrosine (2) and then to melanin (3). A recessive gene in double quantity only allows the conversion of phenylalanine to tyrosine (F) resulting in an albino with a lack of pigment in the skin, hair, and eyes.*

**adhesion** In medicine, fibrous band of connective tissue developing at a site of inflammation or damage; it may bind together adjacent tissues, such as loops of intestine, occasionally causing obstruction. Most adhesions result from inflammation or surgery.

**agglutination** Clumping of BACTERIA or ERYTHROCYTES by antibodies that react with ANTIGENS on the cell surface.

**AIDS** *See* ACQUIRED IMMUNE DEFICIENCY SYNDROME (AIDS)

**albino** Person or animal with a rare hereditary absence of pigment from the skin, hair, and eyes. The hair is white and the skin and eyes are pink because, in the absence of pigment, the blood vessels are visible. The eyes are abnormally sensitive to light and vision is often poor.

**albumin** (albumen) Type of water-soluble protein occurring in animal tissues and fluids. Principal forms are egg albumin (egg white), milk albumin, and blood albumin. In a healthy human, it comprises about 5% of the body's total weight. It is composed of a colorless, transparent fluid called PLASMA in which are suspended microscopic ERYTHROCYTES (red blood cells), LEUKOCYTES (white blood cells), and PLATELETS.

**allergy** Disorder in which the body mounts a hypersensitive reaction to one or more substances (allergens) not normally considered harmful. Typical allergic reactions include sneezing (HAY FEVER), 'wheezing' and difficulty in breathing (ASTHMA), and skin eruptions and itching (ECZEMA). A tendency to allergic reactions is often hereditary.

**Alzheimer's disease** Degenerative condition characterized by memory loss and progressive mental impairment; it is the most common cause of DEMENTIA. Sometimes seen in middle age, Alzheimer's becomes increasingly common with advancing age. Many factors have been implicated, but the precise cause is unknown.

**American Medical Association (AMA)** US federation of 54 state and territorial medical associations, founded in 1847. The AMA develops programs to provide scientific information for the profession and health-education materials for the public. In 2003, more than 260,000 physicians were members of the AMA.

**amphetamine** DRUG that stimulate the central nervous system. These drugs (also known as 'pep pills' or 'speed') can lead to drug abuse and dependence. They can induce a temporary sense of well-being, often followed by fatigue and depression. *See also* ADDICTION

**amylase** Digestive enzyme secreted by the salivary glands (salivary amylase) and the pancreas (pancreatic amylase). It aids digestion by breaking down starch into maltose (a disaccharide) and then glucose (a monosaccharide).

**anabolic steroid** Any of a group of hormones that stimulate the growth of tissue. Synthetic versions are used in medicine to treat OSTEOPOROSIS and some types of ANEMIA; they may also be prescribed to aid weight gain in severely ill or elderly patients. These drugs are associated with a number of side effects, including acne, fluid retention, liver damage, and masculinization in women. Some athletes abuse anabolic steroids in order to increase muscle bulk.

**analgesic** DRUG that relieves or prevents pain without causing loss of consciousness. It does not cure the cause of the pain, but helps to deaden the sensation. Some analgesics are also NARCOTICS. Many have valuable antiinflammatory properties. Common analgesics include aspirin, codeine, and morphine. *See also* ANESTHESIA

**androgen** General name for male sex hormones, such as TESTOSTERONE.

**anemia** Condition in which there is a shortage of hemoglobin, the oxygen-carrying pigment contained in erythrocytes (red blood cells). Symptoms include weakness, pallor, breathlessness, faintness, palpitations, and lowered resistance to infection. It may be due to a decrease in the production of hemoglobin or red blood cells, excessive destruction of red blood cells, or blood loss. Worldwide, iron deficiency is the commonest cause of anemia.

**anesthesia** State of insensibility or loss of sensation produced by disease or by various anesthetic drugs used during surgical procedures. During **general** (or total) anesthesia, the entire human body becomes insensible and the individual sleeps. In **local** anesthesia only a specific part of the body is rendered insensible, and the patient remains conscious. A general anesthetic may be either an injected drug, such as the BARBITURATE **thiopentone**, used to induce unconsciousness, or an inhalation agent such as **halothane**, which is used to maintain anesthesia for surgery. Local anesthetics, such as **lignocaine**, numb the relevant part of the body by blocking the transmission of impulses through the sensory nerves which supply it. Ether was the first anesthetic in general use from 1846.

► *Alzheimer's disease*
*Computer graphic of a vertical (coronal) slice through the brain of a patient with Alzheimer's disease (left) compared with a normal brain (right). The brain of the Alzheimer's patient has shrunk, due to the degeneration and death of nerve cells. Apart from a decrease in brain volume, the surface of the brain is often more deeply folded. Alzheimer's accounts for most cases of senile dementia. Symptoms include memory loss, disorientation, and delusion. It ultimately leads to death.*

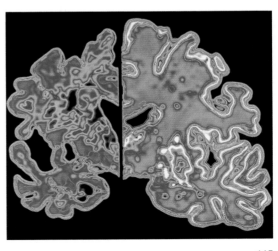

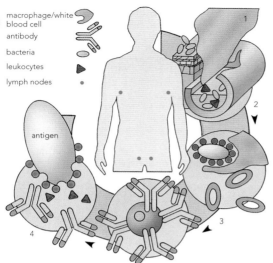

macrophage/white blood cell
antibody
bacteria
leukocytes
lymph nodes

antigen

◀ *Antibody* When bacteria enter the blood stream at the site of a wound (**1**), macrophages (white blood cells) engulf some (**2**) and carry them to the lymph nodes (**3**) where B cells (blue) produce antibodies to bind with the unique antigen proteins on the surface of the invading bacteria (**4**). When the antibodies released into the blood stream attach themselves to the bacteria they stimulate certain leukocytes to destroy them.

**aneurysm** Swelling of an artery caused by damage to, or weakness of, the arterial wall. Aneurysms can occur anywhere in the body, but occur most commonly on the aorta, the body's largest artery, and on the Circle of Willis, a small circle of arteries underneath the base of the brain.

**angina** Pain in the chest due to an insufficient blood supply to the heart, usually associated with diseased coronary arteries. Generally induced by exertion or stress, it is treated with drugs, such as glyceryl trinitrate, or by surgery.

**anorexia nervosa** Abnormal loss of the desire to eat. A pathological condition, it is seen mainly in young women anxious to lose weight. It can result in severe emaciation and in rare cases may be life-threatening. *See also* BULIMIA NERVOSA

**anthrax** Contagious disease, chiefly of livestock, caused by the microbe *Bacillus anthracis.* Human beings can catch anthrax from contact with infected animals or their hides. In 2001, five US citizens died after contact with mail contaminated by anthrax spores, provoking worries of bioterrorism.

**antibiotic** Substance that is capable of stopping the growth of, or destroying, BACTERIA and other microorganisms. Antibiotics are germicides that are safe enough to be eaten or injected into the body. The post-1945 introduction of antibiotics revolutionized medical science, making possible the virtual elimination of once widespread and often fatal diseases, such as TYPHOID FEVER, PLAGUE, and CHOLERA. Some antibiotics are selective – that is, they are effective against specific microorganisms. Those effective against a large number of microorganisms are known as **broad-spectrum** antibiotics. Important antibiotics include PENICILLIN, the first widely used antibiotic, streptomycin and the tetracyclines. Some bacteria have developed ANTIBIOTIC RESISTANCE. *See also* ANTISEPTIC

**antibiotic resistance** Resistance to antibiotic drugs acquired by many BACTERIA and other pathogens (disease-causing microorganisms). Because some bacteria survive while nonresistant strains die, they pass their resistance to their progeny and resistance increases in the population. Some bacteria have the ability to pass the genes for antibiotic resistance to other organisms of different species on plasmids (small lengths of DNA). Spread of resistance is accelerated by routine prescription of antibiotics to humans and unregulated application to farm animals for the purpose of disease prevention rather than cure. An inadequate dose or failure to complete a course of antibiotics increases the chances of resistant microorganisms surviving to breed. This may lead to the return of epidemics of untreatable infectious disease.

**antibody** Protein synthesized in the BLOOD in response to the entry of 'foreign' substances or organisms into the body. Each episode of bacterial or viral infection prompts the production of a specific antibody to fight the disease. After the infection has cleared, the antibody remains in the blood to fight off any future invasion.

**antigen** Substance or organism that induces the production of an ANTIBODY, part of the body's defense mechanism against disease. An antibody reacts specifically with the antigen.

**antihistamine** Any one of certain drugs that counteracts or otherwise prevents the effects of histamine, a natural substance released by the body in response to injury, or more often as part of an allergic reaction. Histamine can produce symptoms such as sneezing, running nose, and burning eyes. See also HAY FEVER

**antiseptic** Chemicals that destroy or stop the growth of many microorganisms. Antiseptics are weak germicides that can be used on the skin. The English surgeon Joseph Lister pioneered the use of antiseptics in 1867. One commonly used is alcohol. See also ANTIBIOTIC

**antitoxin** ANTIBODY produced by the body in response to a TOXIN. It is specific in action and neutralizes the toxin. Antitoxin sera are used to treat and prevent bacterial diseases such as TETANUS and DIPHTHERIA.

**aphasia** Group of disorders of language arising from disease of or damage to the brain. In aphasia, a person has problems formulating or comprehending speech and difficulty in reading and writing.

**appendicitis** Inflammation of the appendix caused by obstruction and infection. Symptoms include severe pain in the central abdomen, nausea, and vomiting. Acute appendicitis is generally treated by surgery. A ruptured appendix can cause PERITONITIS and even death.

**arteriosclerosis** General term for degenerative diseases of the arteries, in particular **atherosclerosis** (hardening of the arteries). It is caused by deposits of fatty materials and scar tissue on the artery walls, which narrow the channel and restrict blood flow, causing an increased risk of heart disease,

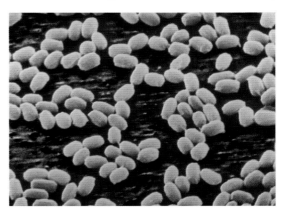

▶ *Anthrax False-color SEM of spores of the rod-shaped bacteria Bacillus anthracis, the causative agent of anthrax in farm animals. The disease is transmitted to man by contact with infected animal hair, hides, or excrement. The bacilli attack either the lungs, causing pneumonia, or the skin, producing severe ulceration (malignant pustule). In the body the bacteria appear singly or in pairs, but when cultured they link up, as seen here. Magnification: ×2,825.*

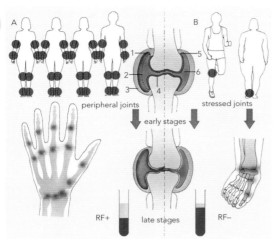

◀ **Arthritis** *In rheumatoid arthritis (**A**) the synovial membrane (**1**) becomes inflamed and thickened and produces increased synovial fluid within the joint (**2**). The capsule and surrounding tissues (**3**) become inflamed, while joint cartilage is damaged (**4**). Peripheral joints are usually affected. Blood tests show rheumatoid factor. Osteoarthritis (**B**), a degenerative disease, involves thinning of cartilage (**5**), loss of joint space (**6**), and bone damage (**7**). Heavily used or weight-bearing joints are affected. Blood tests are normal.*

stroke, or gangrene. Evidence suggests that predisposition to the disease is hereditary. Risk factors include smoking, inactivity, obesity, and a diet rich in animal fats and refined sugar. Treatment is by drugs and, in some instances, surgery is necessary to replace a diseased length of artery.

**arthritis** Inflammation of the joints, with pain and restricted mobility. The most common forms are osteoarthritis and rheumatoid arthritis. OSTEOARTHRITIS, common among the elderly, occurs with erosion of joint cartilage and degenerative changes in the underlying bone. It is treated with analgesics and anti-inflammatories and, in some cases (especially a diseased hip), by joint replacement surgery. **Rheumatoid arthritis**, more common in women, is generally more disabling. It is an autoimmune disease which may disappear of its own accord but is usually slowly progressive. Treatment includes analgesics to relieve pain. The most severe cases may need to be treated with CORTISONE injections, drugs to suppress immune activity, or joint replacement surgery. See also RHEUMATISM

**artificial insemination** Method of inducing PREGNANCY without sexual intercourse by injecting sperm into the female genital tract. Used extensively in livestock farm-

ing, artificial insemination allows proven sires to breed with many females at low cost. See also IN VITRO FERTILIZATION (IVF)

**asthma** Disorder of the respiratory system in which the bronchi (air passages) of the lungs go into spasm, making breathing difficult. It can be triggered by infection, air pollution, allergy, certain drugs, exertion, or emotional stress. **Allergic** asthma may be treated by injections aimed at lessening sensitivity to specific allergens. Otherwise treatment is with bronchodilators to relax the bronchial muscles and ease breathing. In severe asthma, inhaled steroids may be given. Children often outgrow asthma, while some people suddenly acquire the disease in middle age. Air pollution is increasing the number of asthma sufferers. See also BRONCHITIS; EMPHYSEMA

**astigmatism** Defect of vision in which the curvature of the lens differs from one perpendicular plane to another. It can be compensated for by corrective lenses.

**ataxia** Condition where muscles are uncoordinated. It results in clumsiness, irregular and uncontrolled movements, and difficulties with speech. It may be caused by physical injury to the brain or nervous system, by a STROKE, or by disease.

**atrophy** Shrinking or wastage of tissues or organs. It may be associated with disease, malnutrition, or, in the case of muscle atrophy, with disuse.

**atropine** Poisonous alkaloid drug ($C_{17}H_{23}NO_3N$) obtained from certain plants such as *Atropa belladonna* (deadly nightshade). Atropine is used medicinally to regularize the heartbeat during anesthesia, to dilate the pupil of the eye, and to treat motion sickness.

**autism** Disorder, usually first appearing in early childhood, characterized by a withdrawal from social behavior, communication difficulties, and ritualistic behavior. Autistic people usually have difficulty understanding themselves or others as agents with varying beliefs and desires. The causes of autism may originate in genetics, brain damage, or psychology. **Asperger's syndrome** commonly refers to autistic people who are more able, who have better communication skills, and who have greater social contact. It affects about one child in 2500. Boys are affected four times as often as girls are.

**autoimmune disease** Any one of a group of disorders caused by the body's production of antibodies which attack the body's own tissues. One example of such an autoimmune disease is systemic LUPUS ERYTHEMATOSUS (SLE), an inflammation of the connective tissue occurring most often in young women. The occasional presence of so-called autoantibodies in an individual does not necessarily indicate autoimmune disease.

**B**

**Ayurveda** System of medicine practiced by the ancient Hindus and derived from the Vedas (ancient, most sacred writings of Hinduism). It is still practiced in India.

**bacillus** Genus of rodlike BACTERIA that is present everywhere in the air and soil. One example of a species that is pathogenic in man is *Bacillus anthracis*, which causes ANTHRAX.

▶ *Bacteria occur in spherical forms called cocci (**A**), rodlike bacilli (**B**), and spiral spiralla (**C**). Cocci occur in clumps called staphylococci (**1**), pairs called diplococci (**2**), or chains called streptococci (**3**). Unlike cocci, bacilli are mobile; some are termed peritrichous and use flagella (**4**) to swim, while other monotrichous forms use a single flagellum (**5**). Bacilli can also form spores (6) to survive harsh conditions. Spiralla may be either cork screw-shaped spirochaetes like Leptospira (**7**), or less coiled and flagellated, such as Spirillum (**8**)*

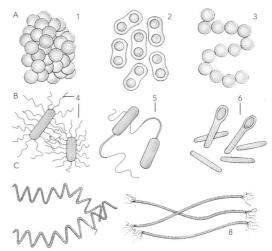

**B**

**bacteria** Simple, unicellular, microscopic organisms. They lack a clearly defined nucleus and most are without chlorophyll. Many are motile, swimming by means of whiplike flagella. In adverse conditions, many can remain dormant inside highly resistant spores with thick protective coverings. Although pathogenic bacteria are a major cause of human disease, many bacteria are harmless or even beneficial to humans by providing an important link in food chains, by decomposing plant and animal tissue, or converting free nitrogen and sulfur into amino acids and other compounds that plants and animals can use. Bacteria belong to the kingdom Prokaryotae. *See picture,* page 119

**bacteriology** Scientific study of BACTERIA. Bacteria were first observed in the 17th century by Anton van Leeuwenhoek, but it was not until the researches of Louis Pasteur and Robert Koch in the mid-19th century that bacteriology was established as a scientific discipline.

**bacteriophage** VIRUS that lives on and infects BACTERIA. It has a protein head containing a core of DNA and a protein tail. Discovered in 1915, it is important in the study of genetics.

**barbiturate** DRUG used as a sedative or to induce sleep. Highly addictive and dangerous in high doses, or in combination with other drugs such as alcohol or tranquilizers, most barbiturates are no longer prescribed. **Short-acting** barbiturates are used in surgery to induce general ANESTHESIA. **Long-acting** formulations are prescribed for epilepsy.

**basal metabolic rate (BMR)** Minimum amount of energy required by the body to sustain basic life processes, including breathing, circulation, and tissue repair. It is calculated by measuring oxygen consumption. Metabolic rate increases above BMR during physical activity or fever, or under the influence of some drugs (including caffeine). It falls below the basal metabolic rate during sleep, general ANESTHESIA, or starvation. BMR is highest in children and decreases with age. *See also* METABOLISM

**BCG** (**B**acille **C**almette **G**uérin) Vaccine against tuberculosis. It was discovered by the French bacteriologists Albert Calmette (1863–1933) and Camille Guérin (1872–1961).

**Bell's palsy** Paralysis of a facial nerve causing weakness of the muscles on one side of the face. The condition, which may be due to viral infection, usually disappears spontaneously, or may be treated with drugs or, rarely, surgery.

◀ *Beta blocker Polarized light micrograph of crystals of the beta-blocker drug Propranolol (Propranolol hydrochloride). Beta-blocker drugs selectively block receptors of the sympathetic nervous system in heart muscle, in the process decreasing the activity of the heart. Propranolol may also be used to treat an overactive thyroid gland, and to prevent migraine. Magnification: ×312.*

**B**

**bends** (decompression sickness) Syndrome, mostly seen in divers, featuring pain in the joints, dizziness, nausea, and paralysis. It is caused by the release of nitrogen into the tissues and blood. This occurs if there is a too rapid a return to normal atmospheric pressure after a period of breathing high-pressure air (when the body absorbs more nitrogen). Treatment involves gradual decompression in a hyperbaric chamber.

**benzodiazepine** Any of a group of mood-altering drugs, such as librium and valium, that are used primarily to treat severe anxiety or insomnia. They intervene in the transmission of nerve signals in the central nervous system, and were originally developed as muscle relaxants. Today, they are the most widely prescribed TRANQUILIZERS. *See* **nervous system**, pages 34–55

**beriberi** Disease caused by a deficiency of vitamin $B_1$ (thiamine) and other vitamins in the diet. Symptoms include weakness, edema (waterlogging of the tissues), and degeneration of nerves. The disease is rare in the developed world.

**beta blocker** Any of a class of DRUGS that block impulses to beta nerve receptors in various tissues throughout the body, including the heart, airways, and peripheral arteries. These drugs are mainly prescribed to regulate the heartbeat, reduce blood pressure, and relieve ANGINA. They are also being used in an increasingly wide range of other conditions, including GLAUCOMA, liver disease, thyrotoxicosis, MIGRAINE, and anxiety states. Beta blockers are not suitable for patients with asthma or severe lung disease.

**bile** Bitter yellow, brown, or green alkaline fluid, secreted by the liver and stored in the gall bladder. Important in digestion, it enters the duodenum via the bile duct. The bile salts it contains emulsify fats (allowing easier digestion and absorption) and neutralize stomach acids. *See* pages 89–93

**bioengineering** Application of engineering techniques to medical and biological problems such as devices to aid or replace defective or inadequate body organs, as in the production of artificial limbs and hearing aids.

**biofeedback** In alternative medicine, the use of monitoring systems to provide information about body processes to enable them to be controlled voluntarily. By observing data on events that are normally involuntary, such as breathing and the heartbeat, many people learn to gain control over them to some extent in order to improve their sense of well-being. The technique has proved helpful in a number of conditions, including MIGRAINE and HYPERTENSION.

**biopsy** Removal of a small piece of tissue from a patient for examination for evidence of disease. An example is the CERVICAL SMEAR (pap test), performed in order to screen for possible precancerous changes that can lead to cervical cancer.

**biotechnology** Use of biological processes for medical, industrial, or manufacturing purposes. Humans have long used yeast for brewing and bacteria for products such as cheese and yoghurt. Biotechnology now enjoys a wider application. By growing microorganisms in the laboratory, new drugs and chemicals are produced. Genetic engineering techniques of cloning, splicing, and mixing genes facilitate, for example, the growing of crops outside their normal environment, and vaccines that fight specific diseases. Hormones are also produced, such as insulin for treating diabetes.

**birth, Caesarean** Delivery of a baby by a surgical incision made through the abdomen and uterus of the mother. It is carried out for various medical reasons; the mother usually recovers quickly, without complications. The procedure is named for Julius Caesar (100–44 BC), who is reputed to have been born this way.

**birth control** *See* CONTRACEPTION

**Black Death** (1347–52) Pandemic of PLAGUE, both bubonic and pneumonic, which killed

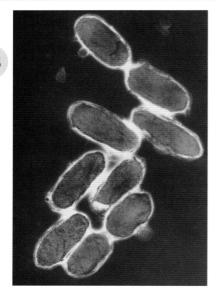

◀ *Black Death* False-color TEM *of* Yersinia pestis, *the bacterium which causes bubonic plague (the Black Death of the Middle Ages). According to the World Health Organization, the bacterium continues to kill about 3,000 people each year*

**blood clotting** Protective mechanism that prevents excessive blood loss after injury. A mesh of tight fibers (of insoluble FIBRIN) coagulates at the site of injury through a complex series of chemical reactions. This mesh traps blood cells to form a clot, which dries to form a scab. This prevents further loss of blood, and also prevents BACTERIA getting into the wound. Normal clotting takes place within five minutes. The clotting mechanism is impaired in some diseases, such as HEMOPHILIA. *See* page 67

**blood group** Any of the types into which blood is classified according to which ANTIGENS are present on the surface of its red cells. There are four major types: A, B, AB, and O. Each group in the ABO system may also contain the rhesus factor (Rh), in which case it is Rh-positive; otherwise it is Rhnegative. Such typing is essential before BLOOD TRANSFUSION, since using blood of the wrong group may produce a dangerous or even fatal reaction. *See* page 67

**blood poisoning** (septicemia) Presence in the blood of BACTERIA or their toxins in sufficient quantity to cause illness. Symptoms include chills and fever, sweating, and collapse. It is most often seen in people who are already in some way vulnerable, such as the young or old, the critically ill or injured, or those whose immune systems have been suppressed.

**blood pressure** Force exerted by circulating BLOOD on the walls of blood vessels due to the pumping action of the heart. This is measured, using a gauge known as a SPHYGMOMANOMETER. It is greatest when the heart contracts, and lowest when it relaxes. High blood pressure is associated with an increased risk of heart attacks and STROKES. Abnormally low blood pressure is mostly seen in people in shock or following excessive loss of fluid or blood.

about one-third of the population of Europe (200 million people) in six years. The plague reached East Asia in the mid-1330s and West Asia a decade later. It was first carried to Mediterranean ports from the Crimea and spread rapidly throughout Europe, carried by fleas infesting rats. Plague recurred less severely in 1361 and other years, until the 18th century.

**blindness** Severe impairment or absence of sight. It may be due to heredity, accident, disease, or old age. Worldwide, the most common cause of blindness is TRACHOMA. In developed countries, blindness is most often due to severe DIABETES, GLAUCOMA, CATARACT, or degenerative changes associated with aging.

**blood** Fluid circulating in the body that transports oxygen and nutrients to all the cells and removes wastes such as carbon dioxide. In a healthy human, it constitutes *c*.5% of the body's total weight; by volume, it comprises *c*.9.7 pints (5.5 liters). It is composed of a colorless, transparent fluid called PLASMA in which are suspended microscopic ERYTHROCYTES, LEUKOCYTES, and PLATELETS. *See* pages 64–66

**blood transfusion** Transfer of blood or a component of blood from one body to another to make up for a deficiency. This is possible only if the BLOOD GROUPS of the donor and recipient are compatible. It is often done to counteract life-threatening SHOCK following excessive loss of blood. Donated blood is scrutinized before use for readily transmissible diseases such as HEPATITIS B and ACQUIRED IMMUNE DEFICIENCY SYNDROME (AIDS).

**boil** (furuncle) Small, pus-filled swelling on the skin, often formed around a hair follicle or sebaceous gland. Most boils are caused by infection from a bacterium (STAPHYLOCOCCUS).

**boric acid** (orthoboric acid) Soft, white crystalline solid ($H_3BO_3$) that occurs naturally in certain volcanic hot springs. It is used as a metallurgical flux, preservative, antiseptic, and an insecticide for pests.

**botulism** Rare, but potentially lethal, form of food poisoning caused by a toxin produced by the bacterium *Clostridium botulinum*. The toxin attacks the nervous system, causing paralysis and cessation of breathing. The most likely source of botulism is imperfectly canned meat. Botulinum toxin (**Botox**) is used medicinally as a treatment for some neuromuscular disorders and in plastic surgery.

**bovine spongiform encephalopathy (BSE)** In cattle, degeneration of the brain caused by infectious particles or PRIONS, which may be transmitted by feeding infected meat. It is also known as 'mad-cow disease.' From 1986 to 2000, more than 4.5 million BSE-infected cattle were slaughtered in the United Kingdom. *See also* CREUTZFELDT-JAKOB DISEASE (CJD)

**brain damage** Result of any harm done to brain tissue causing the death of nerve cells. It may arise from a number of causes, such as oxygen deprivation, brain or other disease, or head injury. The nature and extent of damage varies. Sudden failure of the oxygen supply to the brain may result in widespread (global) damage, whereas a blow to the head may affect only one part of the brain (local damage). Common effects of brain damage include weakness of one or more limbs, impaired balance, memory loss, and personality change; epilepsy may develop. While brain cells do not regenerate, there is some hope of improvement following mild to moderate brain damage, especially with skilled rehabilitation. The survivor of major brain injury is likely to remain severely disabled, possibly even permanently unconscious.

**breathing** Process by which air is taken into and expelled from the lungs for the

▶ *Blood clot Colored SEM of a blood clot. Red blood cells are seen trapped in a web of white fibrin threads made of an insoluble protein. Small platelets (green) have also become enmeshed. When activated, platelets become spiky (as here), clump together, and release chemicals into the blood. These factors cause fibrin threads to form. A white blood cell is also trapped (yellow). A clot that forms inside a blood vessel may lead to a heart attack if it blocks a heart artery. Magnification: ×4,200.*

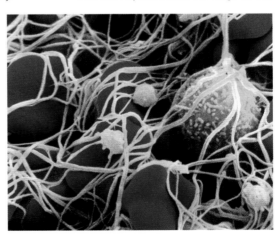

**B**

purpose of gas exchange. During inhalation, the intercostal muscles raise the ribs, increasing the volume of the thorax and drawing air into the lungs. During exhalation, the ribs are lowered, and air is forced out through the nose, and sometimes also the mouth. *See* pages 78–79

**British Medical Association (BMA)** UK professional body founded in 1832; 66% of all doctors in Britain are members. The BMA was set up in order to advance the medical sciences. Since the establishment of the UK NATIONAL HEALTH SERVICE (NHS) in 1948, it has also negotiated pay and conditions for hospital doctors and general practitioners.

**bronchitis** Inflammation of the bronchial tubes, most often caused by a viral infection such as the common cold or influenza but exacerbated by environmental pollutants. Symptoms include coughing and the production of large quantities of mucus. It can be acute (sudden and short-lived) or chronic (persistent), especially in those who smoke.

**BSE** *See* BOVINE SPONGIFORM ENCEPHALOPATHY

**bubonic plague** *See* PLAGUE

**bulimia nervosa** Eating disorder that takes the form of compulsive eating, then purging by induced vomiting or the use of a LAXATIVE or DIURETIC. Confined mainly to girls and women, the disorder most often results from an underlying psychological problem. An obsession with body image may be reinforced by Western media stereotypes of 'slim as beautiful.' *See also* ANOREXIA NERVOSA

**burn** Injury caused by exposure to flames, scalding liquids, caustic chemicals, acids, electric current, or ionizing radiation. Its severity depends on the extent of skin loss and the depth of tissue damage. A **superficial** burn, involving only the epidermis (the outer later of the skin, without blood vessels), causes redness, swelling, and pain; it heals within a few days. A **partial thickness** burn (epidermis and dermis) causes intense pain, with mottling and

blistering of the skin; it takes a couple of weeks to heal. In a **full thickness** burn, involving both the skin and the underlying flesh, there is charring, and the damaged flesh looks dry and leathery; however there is no pain because the nerve endings have been destroyed. Such a burn, serious in itself, is associated with life-threatening complications, including dehydration and infection. Treatment includes fluid replacement and antibiotics; skin grafting may be necessary.

**bursitis** Inflammation of the fluid-filled sac (bursa) surrounding a joint. It is characterized by pain, swelling, and restricted movement. Treatment generally includes rest, heat, and gentle exercise. 'Housemaid's knee', 'tennis elbow', and bunions are common forms of bursitis.

**Caesarean section** *See* BIRTH, CAESAREAN

**caffeine** ($C_8H_{10}N_4O_2$) White, bitter substance that occurs in coffee, tea, and other substances, such as cocoa and ilex plants. It acts as a mild, harmless stimulant and DIURETIC, although an excessive dose can cause insomnia and delirium.

**cancer** Group of diseases featuring the uncontrolled proliferation of cells (TUMOR formation). Malignant (cancerous) cells spread (metastasize) from their original site to other parts of the body. There are many different cancers. Known causative agents (carcinogens) include smoking, certain industrial chemicals, asbestos dust, and radioactivity. Viruses are implicated in the causation of some cancers. Some people have a genetic tendency toward particular types of cancer. Treatments include surgery, chemotherapy with cell-destroying drugs and radiotherapy (or sometimes a combination of all three). Early diagnosis holds out the best chance of successful treatment.

**carbohydrate** Organic compound of carbon, hydrogen, and oxygen. The simplest carbohydrates are sugars. Glucose and fructose are monosaccharides, naturally occurring sugars; they have the same formula ($C_6H_{12}O_6$) but different structures. One molecule of each combines with the loss of water to make sucrose ($C_{12}H_{22}O_{11}$), a disaccharide. Starch and cellulose are polysaccharides, carbohydrates consisting of hundreds of glucose molecules linked together.

**carcinoma** Form of CANCER arising from the epithelial cells present in skin and the membranes lining the internal organs. It is a malignant growth that can increase and invade surrounding tissues, giving rise to further metastases (secondary cancers).

There are two main types of carcinoma: basal cell (or rodent ulcer) and squamous cell. Both are related to sunlight exposure. See also MELANOMA; SARCOMA

**cardiology** Branch of medicine that deals with the diagnosis and treatment of the diseases and disorders of the heart and vascular system.

**castration** Removal of the sexual glands (testes or ovaries) from an animal or human. In human beings, removal of the testes has been used as a form of punishment, a way of sexually incapacitating slaves to produce eunuchs (who were often left in charge of harems), a way of artificially creating soprano voices (castrato), and as a method of stopping the spread of cancer. The procedure can also make animals tamer.

**catabolism** See METABOLISM

**cataract** Opacity in the lens of an eye, causing blurring of vision. Most cases are due to degenerative changes in old age but it can also be congenital, the result of damage to

C

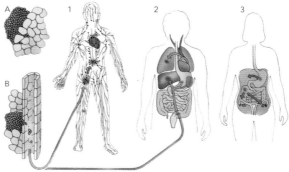

▶ *Cancer can spread in two ways. First, by direct growth into adjacent tissues, called 'direct extension' (A), when cancer cells penetrate into bone, soft connective tissue, and the walls of veins and lymphatic vessels. Second. by 'metastasis' (B), when a cancer cell separates from its tumor and is transported to another part of the body. After the tumor has grown to some size, cancer cells or small groups of cells enter a blood or lymph vessel through the vessel wall (1). They travel through the vessel until they hit a barrier, such as a lymph node, where additional tumors may develop, before releasing more cells which* may develop on other lymph nodes. Such cancers, usually carcinomas, may also invade the blood stream and establish more distant secondary growths. Another type of cancer, sarcomas, tend to spread via venous blood vessels establishing tumors in the lungs, gastrointestinal tract, or the genito-urinary tract (2). In abdominal cancers, metastases may also arise through travel across body cavities, such as the perotoneal, oral, or pleural cavities (3).

C

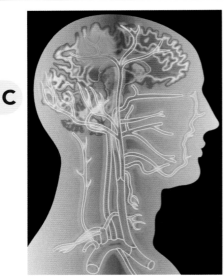

the lens, or some metabolic disorder such as diabetes. Treatment is by removal of the cataract and implanting an artificial lens.

**cephalosporin** Class of ANTIBIOTIC drugs derived from fungi of the genus *Cephalosporium*. Similar to PENICILLIN, they are effective against a wide spectrum of BACTERIA, including some which have become resistant to penicillin.

**cerebral hemorrhage** Form of STROKE in which there is bleeding from a blood vessel in the brain into the surrounding tissue. It is usually caused by ARTERIOSCLEROSIS and high blood pressure. Symptoms may vary from temporary numbness and weakness down one side of the body to deep coma. A major hemorrhage may be fatal. There is a high risk of repeated strokes.

**cerebral hemispheres** Lateral halves of the cerebrum, the largest parts of the brain and the sites of higher thought. Due to the crossing over of nerve fibers from one cerebral hemisphere to the other, the right side controls most of the movements and sensation on the left side of the body, and vice-versa. Damage to cerebral hemispheres can produce personality changes.

◀ *Cerebral hemorrhage Computer artwork of the blood vessels in the human head, showing an intracerebral brain hemorrhage which may lead to a cerebrovascular accident, or stroke. The large orange area at top left is a blood-filled area, or brain hemorrhage. This can be caused by a head injury or by the spontaneous rupture of a blood vessel. The carotid arteries carry blood from the body's main artery, the aorta (seen at bottom center) to the head. Stroke is a major cause of death in developed countries. If patients survive, damage to the brain can cause varied symptoms, such as weakness or paralysis. Brain hemorrhages divide into four main categories – extradural, subdural, subarachnoid, and intracerebral – according to the site of the bleeding.*

**cerebral palsy** Disorder mainly of movement and coordination caused by damage to the brain during or soon after birth. It may feature muscular spasm and weakness, lack of coordination, and impaired movement or paralysis and deformities of the limbs. Intelligence is not necessarily affected. The condition may result from a number of causes, such as faulty development, oxygen deprivation, birth injury, hemorrhage, or infection.

**cerebrospinal fluid** Clear liquid that cushions the brain and spinal cord, giving some protection against shock. It is found between the two innermost meninges (membranes) in the four ventricles of the brain, and in the central canal of the spinal cord. A small quantity of the fluid can be withdrawn by lumbar puncture to aid diagnosis of some brain diseases.

**cervical smear** (pap test) Test for CANCER of the cervix, established by George Papanicolan. In this diagnostic procedure, a small sample of tissue is removed from the cervix and examined under a microscope for the presence of abnormal, precancerous cells. Treatment in the early stages of cervical cancer can prevent the disease from developing.

**chemotherapy** Treatment of a disease (usually cancer) by a combination of chemical substances, or drugs, that kill or impair disease-producing organisms in the body. Specific drug treatment was introduced in the early 1900s by Paul Ehrlich.

**chickenpox** (varicella) Infectious disease of childhood caused by a virus of the HER-PES group. After an incubation period of two to three weeks, a fever develops and red spots (which later develop into blisters) appear on the trunk, face, and limbs. Recovery is usual within a week, although the possibility of contagion remains until the last scab has been shed.

**childbirth** *See* LABOR

**chiropractic** Nonorthodox medical practice based on the theory that the nervous system integrates all of the body's functions, including defense against disease. Chiropractors aim to remove nerve interference by manipulations of the affected musculo skeletal parts, particularly in the spinal region.

**chlamydia** Small, viruslike BACTERIA that live as PARASITES in animals and cause disease.

One strain, *C. trachomatis*, causes TRACHOMA and is also a major cause of pelvic inflammatory disease (PID) in women. *C. psittaci* causes PSITTACOSIS. Chlamydial infection is the most common SEXUALLY TRANSMITTED DISEASE (STD) in many developed countries.

**chloroform** (trichloromethane) Colorless, volatile, sweet-smelling liquid ($CHCl_3$) prepared by the chlorination of methane. Formerly a major anesthetic, it is used in the manufacture of fluorocarbons, in cough medicines, for insect bites, and as a solvent. Properties: sp. gr. 1.48; m.p. −82.3°F (−63.5°C); b.p. 142.2°F (61.2°C).

**cholera** Infectious disease caused by the bacterium *Vibrio cholerae*, transmitted in contaminated water. Cholera, prevalent in many tropical regions, produces almost continuous, watery diarrhea often accompanied by vomiting and muscle cramps, and leads to severe dehydration. Untreated it can be fatal, but proper treatment, including fluid replacement and antibiotics, result in a high recovery rate. There is a vaccine.

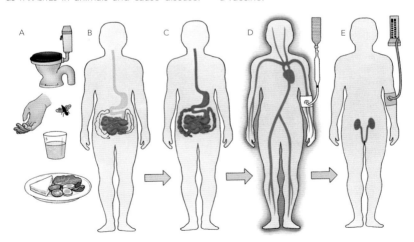

▲ *Cholera* is one of the many diseases that spread because of poor hygiene. Human feces containing cholera germs contaminate food and water (*A*). The victim ingests them and they incubate in his small intestine (*B*). They then produce severe, often fatal diarrhea (*C*). The patient may survive if his gross dehydration can be corrected with intravenous fluid (*D*). If such therapy is undertaken promptly (and if no kidney damage has occurred) then full recovery should take place (*E*).

**C**

**cholesterol** White, fatty STEROID, occurring in large concentrations in the brain, spinal cord, and liver. It is synthesized in the liver, intestines, and skin, and is an intermediate in the synthesis of vitamin D and many hormones. GALLSTONES are composed mainly of cholesterol. Meat-rich diets may produce high cholesterol in blood vessels, and can lead to atherosclerosis. *See also* ARTERIOSCLEROSIS

**chromosome** Structure carrying genetic information, found only in the cell nucleus of eukaryotes. Threadlike and composed of DNA, chromosomes carry a specific set of genes. Each species usually has a characteristic number of chromosomes; these occur in pairs, members of which carry identical genes, so that most cells are diploid. Gametes carry a haploid number of chromosomes.

**circadian rhythm** Internal 'clock' mechanism found in most organisms. It normally corresponds with the 24-hour day, relating most obviously to waking and sleeping cycles, but is also involved in other cyclic variations, such as body temperature, hormone levels, metabolism and mental performance.

**circumcision** Operation of removing part or the whole of the foreskin of the penis or of removing the clitoris. Male circumcision is ritual in some groups, notably

Jews and Muslims, and is said to have sanitary benefits. Female circumcision is intended to reduce sexual pleasure and has no medical benefit.

**cirrhosis** Degenerative disease in which there is excessive growth of fibrous tissue in an organ, most often the liver, causing inflammation and scarring. Cirrhosis of the liver may be caused by viral hepatitis, prolonged obstruction of the common bile duct, chronic abuse of alcohol or other drugs, blood disorder, heart failure, or malnutrition.

**cleft palate** Congenital deformity in which there is an opening in the roof of the mouth, causing direct communication between the nasal and mouth cavities. It is often associated with HARELIP and makes normal speech difficult. Usual treatment includes surgical correction, followed by special dental care and speech therapy if necessary.

**clone** Set of organisms obtained from a single original parent either by asexual reproduction or by artificial selection. Clones are genetically identical and may arise naturally from parthenogenesis in animals. Cloning is often used in plant propagation (including tissue culture) to produce new plants from parents with desirable qualities such as high yield. In 1997 scientists in Scotland produced a sheep embryo from a single cell of an adult sheep using nuclear-transfer technology (transfer of a cell nucleus). *See also* GENETIC ENGINEERING

**coca** Shrub native to Colombia and Peru which contains the alkaloid drug COCAINE. Native Americans chew the leaves for pleasure, to quell hunger, and to stimulate the nervous system. The plant has

◀ *Coca Native to regions of South America, the leaves of the coca tree (Erythroxylon coca) have been used for centuries as medicine and to relieve hunger. The leaves are harvested and used in the illegal manufacture of cocaine. The narcotic principle is the alkaloid cocaine, of which the leaves contain about 1 percent.*

yellow-white flowers growing in clusters, and red berries. Height of shrub: c.8ft (2.4m). Family Erythroxylaceae; species *Erythroxylon coca*.

**cocaine** White, crystalline alkaloid extracted from the leaves of the COCA plant. Once used as a local anesthetic, it is now primarily an illegal narcotic with stimulant and hallucinatory effects. It is psychologically habit-forming, and the body does not develop tolerance. Habitual use of cocaine results in physical and nervous deterioration, and subsequent withdrawal results in severe depression. *See also* CRACK

**coccus** Small spherical or spheroid bacterium. Average diameter: 0.5–1.25 micrometers. Some, such as *Streptococcus* and *Staphylococcus*, are common causes of infection.

**codeine** White, crystalline alkaloid extracted from OPIUM by the methylation of MORPHINE, with the properties of weak morphine. It is used in medicine as an analgesic to treat mild to moderate pain, as a cough suppressant, and to treat diarrhea.

**cold, common** Minor disease of the upper respiratory tract caused by viral infection. Symptoms include inflammation of the nose, headache, sore throat, and a cough. A cold usually disappears within a few days. Fever-reducing and pain-relieving drugs, as well as decongestants, may relieve symptoms. Rest is recommended for heavy colds. Antibiotics may be prescribed where a bacterial infection is also present.

**colic** Severe pain in the abdomen, usually becoming intense, subsiding, and then recurring. Intestinal colic may be associated with obstruction of the intestine or constipation.

**colitis** Inflammation of the lining of the colon, or large intestine, that produces bowel changes, usually diarrhea and cramplike pains. In severe chronic ulcerative colitis, the colon lining ulcerates and bleeds.

**collagen** Protein substance that is the main constituent of bones, tendons, cartilage, connective tissue, and skin. It is made up of inelastic fibers.

**colostomy** Operation to bring the colon out through the wall of the abdomen in order to bypass the lower section of the bowel. An artificial opening is created so that fecal matter is passed into a bag, worn outside the body. The site of the colostomy varies according to the disease's location.

**color blindness** General term for various disorders of color vision. The most common involves red-green vision, a hereditary defect almost exclusively affecting males. Total color blindness (achromatic vision), an inherited disorder in which the person sees only black, white and gray, is very rare. *See page 53*

**coma** State of unconsciousness brought about by head injury, brain disease, drugs, or lack of blood supply to the brain.

**congenital disorder** Abnormal condition present from birth caused by faulty development, infection, or the mother's exposure to drugs or other toxic substances during pregnancy. SPINA BIFIDA is such a condition.

**contact lens** Lens worn on the cornea of the eye to aid defective vision. Invented in 1887, lenses were initially made of glass. Modern contact lenses, developed (1948) by Kevin Tuohy, are made of plastic. **Hard** (corneal) lenses cover the pupil and part of the cornea. They are usually gas-permeable (allowing oxygen to reach the cornea). **Soft** (hydrophilic) lenses cover the whole cornea and are hydrated in saline solution.

**contraception** (birth control) Use of devices or techniques to prevent PREGNANCY. The PILL is a hormone preparation that prevents the release of an egg (ovum) and thickens the cervical mucus. Barrier methods include the male and female condom and the diaphragm. The male **condom** is a latex sheath that covers the penis and

C

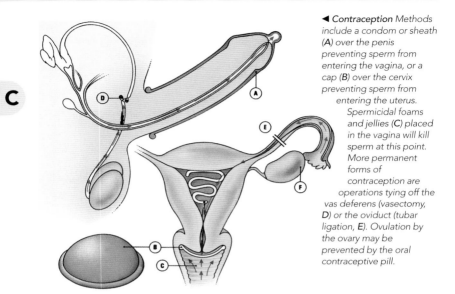

◄ *Contraception Methods include a condom or sheath (A) over the penis preventing sperm from entering the vagina, or a cap (B) over the cervix preventing sperm from entering the uterus. Spermicidal foams and jellies (C) placed in the vagina will kill sperm at this point. More permanent forms of contraception are operations tying off the vas deferens (vasectomy, D) or the oviduct (tubar ligation, E). Ovulation by the ovary may be prevented by the oral contraceptive pill.*

collects the ejaculated semen; the female condom lines the inside of the vagina, preventing any sperm entering the womb. The use of condoms is widely advocated because they protect against some SEXUALLY TRANSMITTED DISEASES (STDs), including ACQUIRED IMMUNE DEFICIENCY SYNDROME (AIDS). Devices, such as the diaphragm or cap, cover the cervix thus preventing sperm entering the womb. Less effective is the 'rhythm method,' which involves the avoidance of sex on days when conception is most likely (when the woman is ovulating). It is not a reliable method as ovulation cannot always be predicted accurately. Emergency contraception, known as the 'morning-after pill,' can be taken up to 72 hours after unprotected sexual intercourse; it prevents the fertilized ovum embedding itself in the womb. It is not suitable to be used regularly.

**convulsion** Intense, involuntary contraction of the muscles, sometimes accompanied by loss of consciousness. A seizure may indicate EPILEPSY although there are other causes, including intoxication, brain abscess, or HYPOGLYCEMIA.

**coronary artery disease** Disease of the coronary blood vessels, particularly the aorta and arteries supplying blood to the heart. *See also* ANGINA; ARTERIOSCLEROSIS

**coronary heart disease** ARTERIOSCLEROSIS of the coronary arteries. It is the most common cause of death in the West. Atheriosclerosis can lead to the formation of a blood clot in one or other of the coronary arteries supplying the heart (**coronary thrombosis**). The patient experiences sudden pain in the chest (ANGINA) and the result may be a HEART ATTACK (**myocardial infarction**), when the flow of blood to the heart is stopped. Smokers are more likely to die suddenly from atheriosclerosis. Evidence suggests that a high intake of polyunsaturated fats can protect against coronary heart disease.

**cortisone** Hormone produced by the cortex of the adrenal glands (on top of the kidneys) and essential for carbohydrate, protein, and fat METABOLISM, kidney function, and disease resistance. Synthetic cortisone is used to treat adrenal insufficiency, rheumatoid arthritis, and other inflammatory diseases and rheumatic fever. *See also* page 57

**cot death** *See* SUDDEN INFANT DEATH SYNDROME (SIDS)

**crack** Street DRUG that is a COCAINE derivative. It is supplied in the form of hard, crystalline lumps, which are heated to produce smoke inhaled for its stimulant effects. It imposes considerable strain on the heart and blood vessels, and may result in heart failure or a stroke. Psychotic episodes may also occur.

**Creutzfeldt-Jakob disease (CJD)** Rare, degenerative brain disease that causes physical deterioration and dementia, usually leading to death within a year of onset. Caused by an abnormal protein (a prion), it is related to scrapie in sheep, and 'mad cow disease' or BOVINE SPONGIFORM ENCEPHALOPATHY (BSE). Typically it affects older people, but in 1996 scientists found a new variant form of CJD (nvCJD) in younger victims. In 1997 research confirmed that the agent responsible for this variant was identical to that of BSE, confirming the link between CJD and the consumption of infected beef. There is currently no known cure.

**croup** Respiratory disorder of children caused by inflammation of the larynx (voice box) and airways. It is triggered by viral infection. Symptoms are a cough, difficulty in breathing, and fever.

**CT scan** (computed tomography) X-ray technique for displaying images of cross-sections through the human body. X-ray sources and detectors slowly move around the patient's body on opposite sides, producing a changing 'view' of an organ. The data from the detectors is processed through a computer to display only the details relating to a specific 'slice' through the body.

**curare** Poisonous resinous extract obtained from various tropical South American plants of the genera *Chondodendron* and *Strychnos*. Most of its active elements are alkaloids. It is used as a muscle relaxant in abdominal surgery and setting fractures.

**cystic fibrosis** Hereditary glandular disease in which the body produces abnormally thick mucus that obstructs the breathing passages, causing chronic lung disease. There is a deficiency of

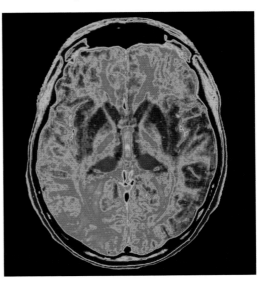

▶ *Creutzfeldt-Jakob disease (CJD) Colored Magnetic Resonance Imaging (MRI) scan of the brain of a 17-year-old male suffering from Creutzfeldt-Jakob disease (CJD: new variant) in 1997. The patient died. The front of the head is at top. In this axial 'slice' through the brain, the folded cerebrum is seen forming two hemispheres. At lower center are two red areas of the thalamus diseased with CJD. It is detected as a 'bilateral signal abnormality' on MRI. CJD destroys nerve cells and causes brain tissue to become spongy. Symptoms include dementia and muscle contractions, and death. There is concern that eating beef from 'mad cows' may cause CJD.*

► *Cystic fibrosis* *Coloured SEM of mucus in a bronchus of the lung of a patient suffering cystic fibrosis. A large clump of mucus is at centre (red). The bronchial airways are lined with hair-like cilia (green, yellow). At top, rounded goblet cells (blue) secrete the mucus that catches inhaled particles. People with cystic fibrosis suffer chronic infection of and mucus production in the lungs.*

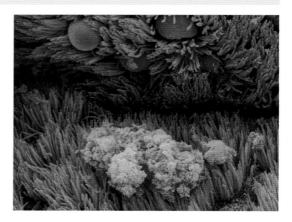

**C**

pancreatic enzymes, an abnormally high salt concentration in the sweat and a general failure to gain weight. The disease is treated with antibiotics, pancreatic enzymes and a high-protein diet. Sufferers must undergo vigorous physiotherapy to keep the chest as clear as possible.

**cystitis** Inflammation of the urinary bladder, usually caused by bacterial infection. It is more common in women. Symptoms include frequent and painful urination, low back pain and slight fever.

**deafness** Partial or total hearing loss. **Conductive** deafness, faulty transmission of sound to the sensory organs, is usually due to infection or inherited abnormalities of the middle ear. **Perceptive** deafness may be hereditary or due to injury or disease of the cochlea in the ear, auditory nerve or hearing centres in the brain. Treatment of deafness ranges from removal of impacted wax to delicate microsurgery. Hearing aids, sign language, and lip-reading are some of the different techniques which help deaf people to communicate. *See* pages 54–5

**death** Cessation of life. In medicine, death has traditionally been pronounced on cessation of the heartbeat. However, modern resuscitation and life-support techniques have enabled the revival of patients whose hearts have stopped. In a tiny minority of cases, while breathing and heartbeat can be maintained artificially, the potential for life is extinct. In this context, death may be pronounced when it is clear that the brain no longer controls vital functions. The issue is highly controversial.

**decompression sickness** *See* BENDS

**dehydration** Removal or loss of water from a substance or tissue. Water molecules can be removed by heat, catalysts or a dehydrating agent such as concentrated sulphuric acid. Dehydration is used in food preservation. In medicine, excessive water loss is often a symptom or result of disease or injury.

**delirium** State of confusion in which a person becomes agitated and incoherent and loses touch with reality; often associated with delusions or HALLUCINATIONS.

Delusion may be seen in various disorders, brain disease, fever, and drug or alcohol intoxication.

**dementia** Deterioration of personality and intellect that can result from disease of or damage to the brain. It is characterized by memory loss, impaired mental processes, personality change, confusion, lack of inhibition and deterioration in personal hygiene. Dementia can occur at any age, although it is more common in the elderly. *See also* ALZHEIMER'S DISEASE

**dengue** Infectious virus disease transmitted by the *Aedes aegypti* mosquito. Occurring in the tropics and some temperate areas, it produces fever, headache and fatigue, followed by severe joint pains, aching muscles, swollen glands, and a rash.

**dentistry** Profession concerned with the care and treatment of the mouth, particularly the teeth and their supporting tissues. As well as general practice, dentistry includes specialities such as oral surgery, periodontics, and orthodontics.

**dermatitis** Inflammation of the skin. In acute form, it produces itching and blisters. In chronic form, it causes thickening, scaling and darkening of the skin. *See also* ECZEMA

**D**

**dermatology** Branch of medicine that deals with the diagnosis and treatment of skin diseases.

**diabetes** Disease characterized by lack of INSULIN needed for sugar METABOLISM. This leads to HYPERGLYCAEMIA and an excess of sugar in the blood. Symptoms include abnormal thirst, over-production of urine, and weight loss. In addition, degenerative changes occur in blood vessels. Untreated, the condition progresses to diabetic coma and death. There are two forms of the disease. **Type**

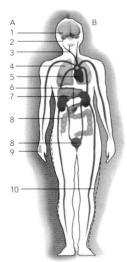 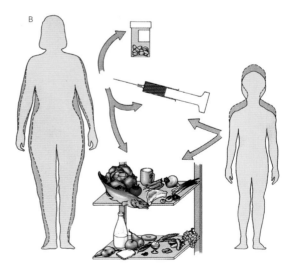

▲ *Diabetes mellitus (A) may cause drowsiness and coma (1), impaired vision (2), dry mouth (3), over-breathing (4), cardiac failure (5), high blood sugar levels (6), fatty liver (7),* kidney and bladder infections *(8),* itchy skin and delayed healing *(9),* and weight loss *(10). Treatment (B) for juveniles includes a varied low-fat diet, usually with* insulin injections. Adults also need a low-fat diet and weight loss, but drugs may help to control the disease. If not, then insulin may be administered.

**D**

the disease. **Type 1** usually begins in childhood and is an autoimmune disease. Those affected owe their survival to insulin injections. **Type 2** diabetes mostly begins in middle age; there is some insulin output but not enough for the body's needs. The disease can be managed with dietary restrictions, tablets to lower blood sugar levels, and insulin injections. Susceptibility to *diabetes mellitus* is inherited and is more common in males.

**dialysis** Process for separating particles from a solution by virtue of differing rates of diffusion through a semipermeable membrane. In an artificial kidney, unwanted molecules of waste products are separated out to purify the blood. Electrodialysis employs a direct electric current to accelerate the process, especially useful for isolating proteins.

**diarrhea** Frequent elimination of loose, watery stools, often accompanied by cramps and stomach pains. It arises from various causes, such as infection, intestinal irritants, or food allergy. Mild attacks can be treated by replacement fluids.

**diet** Range of food and drink consumed by an animal. The human diet falls into five main groups of necessary nutrients: PROTEIN, CARBOHYDRATE, FAT, VITAMIN, and MINERAL. An adult's daily requirement is about one gram of protein for each kilogram of body weight. Beans, fish, eggs, milk, and meat are important protein sources. Carbohydrates (stored as glycogen) and fat are the chief sources of energy and are found in cereals, root vegetables, and sugars. Carbohydrates make up the bulk of most diets. Fats are a concentrated source of energy and aid the absorption of fat-soluble vitamins (vitamins A, D, E, and K). Water and minerals, such as iron, calcium, potassium, and sodium, are also essential.

**diphtheria** Acute infectious disease characterized by the formation of a membrane in the throat which can cause asphyxiation; there is also release of a toxin which can damage the nerves and heart. Caused by a bacterium, *Corynebacterium diphtheriae*,

which often enters through the upper respiratory tract, it is treated with antitoxin and antibiotics.

**diploid** Cell that has its CHROMOSOMES in pairs. Diploids are found in almost all animal cells, except GAMETES which are HAPLOID. In diploids, the chromosomes of each pair carry the same genes.

**disease** Any departure from health, with impaired functioning of the body. Disease may be **acute**, severe symptoms for a short time; **chronic**, lasting a long time; or **recurrent**, returning periodically. There are many types and causes of disease: infectious, caused by harmful BACTERIA or viruses; hereditary and metabolic; growth and development; immune-system diseases; neoplastic (TUMOR-producing); nutritional deficiency; endocrine system diseases; or diseases due to environmental agents, such as lead poisoning. Treatment depends on the cause and course of the disease. It may be **symptomatic** (relieving symptoms, but not necessarily combating a cause) or **specific** (attempting to cure an underlying cause). Disease prevention includes eradication of harmful organisms, VACCINES, public health measures, and routine medical checks.

**diuretic** Drug used to increase the output of urine. It is used to treat raised blood pressure and EDEMA.

**DNA** (deoxyribonucleic acid) Molecule found in all cells, and in some VIRUSES, which is responsible for storing the GENETIC CODE. It consists of two long chains of alternating sugar molecules and phosphate groups linked by nitrogenous bases. The whole molecule is shaped like a twisted rope ladder, with the nitrogenous bases forming the rungs. The sugar is deoxyribose, and the four bases are adenine, cytosine, guanine, and thymine. A base and its associated sugar are known as a **nucleotide**; the whole chain is a polynucleotide chain. The genetic code is stored in terms of the sequence of nucleotides: three nucleotides code for one specific amino acid and a series of them constitute a gene. In eukaryote cells, DNA is stored in chromosomes inside the nucleus. Loops of DNA also occur inside chloroplasts and mitochondria. *See also* RNA

**dopamine** Chemical normally found in the corpus striatum region of the human brain. Insufficient levels are associated with PARKINSON'S DISEASE. Dopamine is a NEUROTRANSMITTER and a precursor in the production of EPINEPHRINE and NOREPINEPHRINE.

**Down's syndrome** Human condition caused by a chromosomal abnormality. It gives rise to varying degrees of mental retardation, decreased life expectancy, and perhaps physical problems such as heart and respiratory disorders. The syndrome was first described by a British physician, J.L.H. Down. It is caused by the presence of an extra copy of chromosome 21, and detected by counting chromosomes in the cells of the fetus during prenatal testing. There is evidence that the risk of having a Down's child increases with maternal age. Originally called 'Mongolism' by Down, this term is now obsolete. *See also* Genes and Inheritance, pages 110–111

**dream** Mental activity associated with the rapid-eye-movement (REM) period of sleep. It is usually a train of thoughts,

scenes and desires expressed in visual images and symbols. On average, a person dreams for a total of 1.5–2 hours in eight hours of sleep. Dream content is often connected with body changes. For centuries, dreams have been regarded as a source of prophecy or visionary insight. In psychoanalysis, patients' dreams are often examined to reveal their latent content.

**D**

**drug** In medicine, any substance used to diagnose, prevent or treat disease, or aid recovery from injury. Although many drugs are still obtained from natural sources, scientists are continually developing synthetic drugs which work on target cells or microorganisms. Such drugs include ANTIBIOTICS. Some drugs interfere in physiological processes, such as anticoagulants which render the blood less prone to clotting. Drugs also may be given to make good some deficiency, such as hormone preparations which compensate for an underactive gland.

**drug addiction** Psychological or physical dependence on a DRUG. **Physical** addiction is often manifested by symptoms of withdrawal. Long-term drug use often produces tolerance. Physical addiction has only been medically proven for NARCOTICS (such as HEROIN), depressants (such

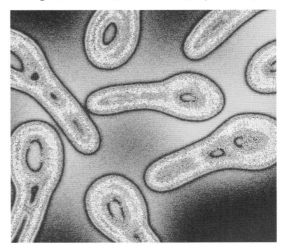

► *Diphtheria Colored TEM of Corynebacterium diphtheriae, the bacteria that causes diphtheria. The bacteria are club-shaped, gram-positive, nonmotile rods. They have a spiral nuclear region (red and orange in center of bacteria) contained within cytoplasm (white speckled). Surrounding each cell is a mucus cell wall (red outlines). Diphtheria is caused by a potent exotoxin released from the bacterium. Symptoms include sore throat, fever, and breathing difficulties. Magnification: x30,000.*

as BARBITURATES or alcohol), and some STIMULANTS (such as NICOTINE). Other drugs, such as hallucinogens or hashish, are not thought to be physically addictive, but can produce PSYCHOSIS or paranoia. Two of the most common addictions are alcohol and nicotine, since these are easily available. Unlike narcotics, alcohol is physically harmful, especially to the brain and liver, and cigarette smoking annually accounts for more than 100,000 premature deaths in the United Kingdom alone. In comparison, addiction to 'hard' drugs (such as HEROIN or COCAINE) is not common, but drug-related crime makes up a significant percentage of crime statistics in many countries.

**dysentery** Infectious disease characterized by DIARRHEA, bleeding, and abdominal cramps. It spreads in contaminated food and water, especially in the tropics. There are two types: **bacillary** dysentery, caused by BACTERIA of the genus *Shigella*; and **amoebic** dysentery, caused by a type of protozoan (unicellular organism). Both forms are treated with antibacterials and fluid replacement.

**dyspepsia** (indigestion) Pain or discomfort in the stomach or abdomen arising from digestive upset.

**ebola** Virus that causes hemorrhagic fever. Thought to have been long present in animals, the ebola virus was first identified in humans during an outbreak in Congo in 1976. It is acquired through contact with contaminated body fluids. Death rates can be as high as 90%.

**ecstasy (MDMA)** (3,4–methylnedioxy-methylamphetamine) AMPHETAMINE-based drug, which raises body temperature and blood pressure by inducing the release of epinephrine and targeting the neurotransmitter serotonin. Users experience short-term feelings of euphoria, rushes of energy, and increased tactility. Withdrawal can involve bouts of depression and INSOMNIA.

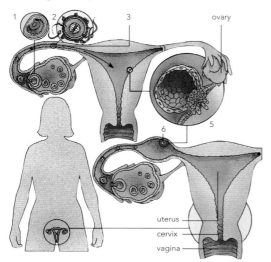

◄ *Ectopic pregnancy* An ectopic or extrauterine pregnancy occurs when the egg, released from the ovary (**1**) and fertilized by the sperm (**2**) in the Fallopian tube (**3**), fails to pass down into the uterus (**4**) where it would normally implant in the wall of the womb (**5**). If the zygote implants in the wall of the Fallopian tube (**6**), there is not enough space or the blood supply to nurture a fetus to full term and the pregnancy usually aborts spontaneously. If not, surgery is necessary to remove the fetus and save the mother.

▶ *Embolism Illustration of a venous embolism (obstruction) caused by a thrombus (clot) lodged in a vein. The clot (red) is seen resting in the cusp of a valve that prevents backflow of blood in the vein. The clot is in a region of disturbed blood flow, notably near the valve. Thrombosis (clot formation) may be caused by disturbed blood flow, together with some condition that increases the natural tendency of the blood to clot.*

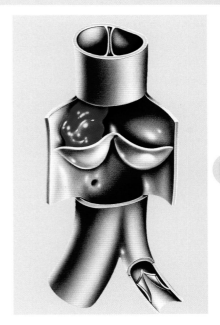

**E**

Since the 1980s the use of ecstasy in tablet form has been widespread in club cultures in Europe and North America. Deaths have resulted from use of the drug – mostly due to dehydration.

**ectopic** Occurrence of a pregnancy outside the uterus, such as in the Fallopian tube. The embryo cannot develop normally, and spontaneous ABORTION (miscarriage) often occurs. If not, urgent surgery is necessary to save the mother from serious hemorrhage. *See also* pages 104–105

**eczema** Inflammatory condition of the skin, a form of DERMATITIS characterized by dryness, itching, rashes, and blister formation. It can be caused by contact with a substance, such as a detergent, to which the skin has been sensitized, or a general ALLERGY. Treatment is usually with a corticosteroid ointment.

**edema** Abnormal accumulation of fluid in the tissues; it may be generalized or confined to one part, such as the ankles. It may be due to heart failure, obstruction of one or more veins, or the increased permeability of the capillary walls.

**electrocardiogram (ECG)** Recording of the electrical activity of the heart, traced on a moving strip of paper by an electrocardiograph. It is used to diagnose heart disease.

**electroconvulsive therapy (ECT)** Treatment of mental disturbance by means of an electric current passed via electrodes to one or both sides of the brain to induce convulsions. Given under anesthesia, it is recommended mainly for depression that has failed to respond to other treatments. It can produce unpleasant side effects, such as confusion, memory loss, and headache. ECT is highly controversial.

**electroencephalogram (EEG)** Recording of electrical activity of the brain. Electrodes are attached to the scalp to pick up the tiny oscillating currents produced by brain activity. It is used mainly in the diagnosis and monitoring of EPILEPSY.

**elephantiasis** Condition in which there is gross swelling of the tissues due to blockage of lymph vessels. It is usually caused by parasitic worms, as in FILARIASIS.

**embolism** Blocking of a blood vessel by an obstruction called an embolus, usually a blood clot, air bubble, or particle of fat. The effects depend on where the embolus lodges; a cerebral embolism causes a STROKE. Treatment is with anticoagulants or surgery. *See also* ARTERIOSCLEROSIS

**emphysema** Accumulation of air in tissues, usually occurring in the lungs (pulmonary emphysema). Pulmonary emphysema, characterized by marked breathlessness, is the

**E**

result of damage to and enlargement of the alveolus (tiny air sacs in the lungs). This type of emphysema is associated with chronic bronchitis and smoking.

**encephalitis** Inflammation of the brain, usually associated with a viral infection; often there is an associated MENINGITIS. Symptoms include fever, headache, lassitude, and intolerance of light; in severe cases there may be sensory and behavioral disturbances, paralysis, convulsions, and even coma. Tests on CEREBROSPINAL FLUID provide a diagnosis of encephalitis.

**endometriosis** Gynecological disorder in which tissue similar to the endometrium (mucous membrane) is found in other parts of the pelvic cavity. It is treated with analgesics, hormone preparations, or surgery.

**endorphin** (endogenous morphine) Naturally occurring peptide NEUROTRANSMITTER found in the pituitary gland that has similar pain-relieving effects as MORPHINE and other derivatives of OPIUM. Endorphins block the sensation of pain by binding to pain-receptor sites. Two endorphins (**enkephalins**) occur in the brain, spinal cord, and gut. See also ANALGESIC

**endoscope** Instrument used to examine the interior of the body. Early types of endoscope, such as the opthalmoscope, were developed in the 19th century. Fiber optics has revolutionized the design of endoscopes. The modern endoscope is a flexible, fiberglass instrument that can be swallowed by a patient, or introduced through a tiny incision in the body. Most endoscopy is for diagnostic purposes, but endoscopes are used in biopsy and MINIMAL ACCESS SURGERY.

**enzyme** (Gk. zymosis, 'fermentation') Protein that functions as a catalyst in biochemical reactions. They remain chemically unaltered in these reactions and so are effective in tiny quantities. Chemical reactions can occur several thousand or million times faster with enzymes than without. They operate within a narrow temperature range, usually 86°F to 104°F (30°C to 40°C) and have optimal pH

ranges. Many enzymes have to be bound to nonprotein molecules to function. These molecules include trace elements (such as metals) and coenzymes (such as vitamins).

**epilepsy** Disorder characterized by abnormal electrical discharges in the brain which provoke seizures. It is seen both in generalized forms, involving the whole of the cerebral cortex (largest section of the brain), or in partial (focal) attacks arising in one small part of the brain. Attacks are often presaged by warning symptoms, the 'aura.' Seizure types vary from the momentary loss of awareness seen in *petit mal* attacks ('absences') to the major convulsions of *grand mal* epilepsy. Attacks may be triggered by a number of factors, including sleep deprivation, flashing lights, or excessive noise. All forms of epilepsy are controlled with anticonvulsant drugs.

**epinephrine** Hormone secreted by the adrenal glands (see p. 57), important in preparing the body's response to stress. It has widespread effects in the body, increasing the strength and rate of heartbeat and the rate and depth of breathing, diverting blood from the skin and digestive system to the heart and muscles, and stimulating the release of glucose from the liver to increase energy supply by promoting increased respiration. Synthetic epinephrine is used medicinally in the resuscitation of patients in shock or following cardiac arrest.

**erythrocyte** Red blood cell, usually diskshaped and without a nucleus. It contains hemoglobin, which combines with oxygen and gives blood its red color. Normal human blood contains an average of 5 million such cells per cu mm of blood.

**estrogen** Female sex hormone. First produced by a girl at PUBERTY, estrogen leads to the development of secondary sexual characteristics: breasts, body hair and redistributed fat. It regulates the MENSTRUAL CYCLE and prepares the uterus for pregnancy. Estrogen is also a constituent of the contraceptive PILL. See pages 100–101

**excretion** Elimination of materials from the body that have been involved in METABOLISM. Such waste materials, particularly nitrogenous wastes, would be toxic if allowed to accumulate. In mammals, these wastes excrete mainly as URINE, and to some extent by sweating. Carbon dioxide excretes through the lungs during breathing. Defecation is not excretion, as feces consist mostly of material that has never been part of the body. *See pages 82–93*

**fainting** (syncope) Temporary loss of consciousness accompanied by general weakness of the muscles. A faint may be preceded by giddiness, nausea, and sweating. Its causes include insufficient flow of blood to the brain and shock.

**family planning** Alternative term for CONTRACEPTION

**famine** Extreme prolonged shortage of food, produced by both natural and man-made causes. If it persists, famine results in widespread starvation and death. Famine is often associated with drought, or alterations in weather patterns, which leads to crop failure and the destruction of livestock. However, warfare and complex political situations resulting in the mismanagement of food resources are equally likely causes.

**fertility drugs** Drugs taken to increase a woman's chances of conception and pregnancy. One cause of female sterility is insufficient secretion of pituitary hormones; this can be treated with human chorionic gonadotropin or clomiphene citrate, although use of the latter can result in multiple births. In cases where fertilization occurs, but where the uterine lining is unable to support the fetus, the hormone PROGESTERONE may be used. Infertility cannot always be corrected with drugs.

**fever** Elevation of the body temperature above normal, that is 98.6°F (37°C). It is mostly caused by bacterial or viral infection and can accompany virtually any infectious disease.

**fibrin** Insoluble, fibrous protein that is essential to BLOOD CLOTTING. Developed in the blood from a soluble protein, fibrinogen, fibrin is laid down at the site of a wound in the form of a mesh, which then dries and hardens so the bleeding stops.

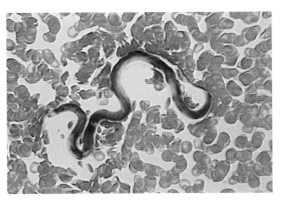

▶ *Filariasis Colored LM of a Loa loa worm on a human blood smear. The worm can be seen among the red blood cells (colored gray). The Loa loa worm is a parasitic nematode that lives under human skin, causing a form of filariasis called loiasis. The worms are spread by bloodsucking Chrysops flies in tropical Africa, and reach 1–3in (3–7cm) when mature. Magnification unknown.*

**F**

**filariasis** Group of tropical diseases caused by infection with a nematode worm, *filaria*. The parasites, which are transmitted by insects, infiltrate the lymph glands, causing swelling and impaired drainage. Drug treatment reduces the symptoms. *See picture*, page 139

**first aid** Action taken by anyone encountering sudden illness or injury in order to save life, mitigate harm, or assist subsequent treatment. Properly administered first aid can save lives and improve the extent and speed of recovery. If the victim appears to have a broken bone or internal injuries, they should not be moved. If the victim is unconscious, their head should be turned to one side to prevent choking. Check that the victim has an open airway – if not, they may have respiratory failure (asphyxia). **Asphyxia** may be caused by the obstruction of air passages, in which case the HEIMLICH MANEUVER is recommended. For asphyxia caused by fumes or gas, such as carbon monoxide, the victim should be moved to a clear atmosphere before administering artificial respiration. The best method of **artifical respiration** is mouth-to-mouth resuscitation. Place the victim on their back, put one hand under the victim's chin and the other on the forehead. Tilt the victim's head back by lifting with the hand under the chin and pressing down on the forehead. If the victim is a child, take a breath, place your mouth over both the nose and mouth, and blow gently. Remove your mouth and listen for air entering the child's lungs. Take a breath and repeat every three seconds until the child begins to breathe. If the victim is an adult, pinch the nostrils shut, take a deep breath, cover the mouth tightly and breathe hard. Repeat this procedure every five seconds. If the victim has swallowed a poisonous substance, identify the **poison** and then call the emergency services. They may recommend inducing vomiting with syrup of ipecac. In the case of animal **bites and stings**, the wound should be cleaned with soap and water before applying antiseptic and a bandage. For poisonous SNAKEBITE, the wound should be cooled with ice to slow down the absorption of poison. A few snake bites require antivenoms. In order to stop the victim **bleeding** severly, apply direct pressure preferably with a sterile dressing. If bleeding continues, apply pressure to the artery that supplies blood to the area. To treat **shock**, place the victim on their back with their legs raised slightly, and place a blanket over the body to prevent heat loss. In the case of superficial and partial thickness **burns**, cold water should be applied before dressing with sterile bandages.

**flu** Abbreviation of INFLUENZA

**follicle-stimulating hormone (FSH)** Hormone produced by the anterior pituitary gland found in the brain of mammals. In females, it regulates ovulation by stimulating the Graafian follicles found in the OVARY to produce eggs (ova). In males, FSH promotes spermatogenesis. FSH is an ingredient in most fertility drugs.

**food poisoning** Acute illness caused by consumption of food that is itself poisonous or which has become poisoned or contaminated with bacteria. Frequently implicated are SALMONELLA bacteria, found in cattle, pigs, poultry, and eggs, and listeria, sometimes found in certain types of cheese. Symptoms include abdominal pain, DIARRHEA, nausea, and vomiting. Treatment includes rest, fluids to prevent dehydration and, possibly, medication. *See also* BOTULISM; GASTROENTERITIS

**frostbite** Freezing of living body tissue in subzero temperatures. Frostbite is an effect of the body's defensive response to intense cold, which is to shut down blood vessels at the extremities in order to preserve warmth at the core of the body. Consequently, it mostly occurs in the face, ears, hands, and feet. In superficial frostbite, the affected part turns white and cold; it can be treated by gentle thawing. In deep frostbite, ice crystals form in the tissues. The flesh hardens and sensation is lost. It requires urgent medical treatment. No attempt should be made at rewarming if there is a risk of refreezing, as this results in the death of body tissue (gangrene).

**gallstone** (cholelithiasis) Hard mass, usually composed of cholesterol and calcium salts, which forms in the gall bladder. Gallstones may cause severe pain or become lodged in the common bile duct, causing obstructive JAUNDICE or cholecystitis. Treatment is by removal of the stones or of the gall bladder.

**gamete** Reproductive sex cell that joins with another sex cell to form a new organism. Female gametes (ova) are usually motionless; male gametes (SPERM) often have a tail (flagellum) enabling them to swim to the OVUM. All gametes are HAPLOID.

**gastric juice** Fluid comprising a mixture of substances, including PEPSIN and hydrochloric acid, secreted by glands of the stomach. Its principal function is to break down proteins into polypeptides during digestion.

▶ *Gene replacement therapy (GRT)*
*GRT is used to treat severe combined immunodeficiency disease (SCID), where the gene that produces adenosine deaminase (ADA) is missing. As ADA is essential for white blood cell production, this renders the body open to infection. Two retroviruses (1) are introduced into the bone marrow. These have the ability to produce RNA from their DNA (2) using a reverse transcriptase enzyme (3). This DNA is then incorporated into the chromosomes (4). When these chromosomes multiply, new viral RNA, viral proteins, and ADA are produced (5). The first two produce more new viruses, while the body uses ADA to produce white blood cells.*

**gastroenteritis** Inflammation of the stomach and intestines causing abdominal pain, diarrhea, and vomiting. It may be caused by infection, food poisoning, or allergy. Severe cases can cause dehydration. Treatment includes fluid replacement.

**gene replacement therapy (GRT)** Medical treatment involving the replacement or alteration of faulty genes by means of GENETIC ENGINEERING. Although first used on humans in 1990, it remains a largely experimental procedure. A healthy gene is packaged into some kind of vector (usually a suitably doctored virus) so that it can be targeted at the affected cells. Initially, gene therapy was restricted to the treatment of hereditary disorders such as CYSTIC FIBROSIS and SICKLE CELL ANEMIA. Research is being conducted into its suitability as a method for the treatment of certain cancers. *See also* Genes and Inheritance, pages 110–111

**genetic code** Arrangement of information stored in genes. It is the ultimate basis of heredity and forms a blueprint for the entire organism. The genetic code is based on the genes that are present, which, in molecular terms, depends on the arrangement of nucleotides in the long molecules of DNA in the cell chromosomes. Each group of three nucleotides specifies, or codes, for an amino acid, or for an action such as start or stop. By specifying

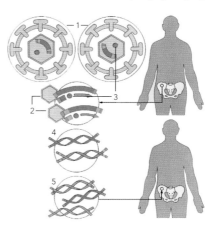

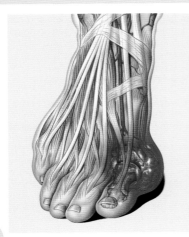

**G**

◀ *Gout Illustration of a foot, showing a tophus (swelling) due to gout. The tophus is seen at the base of the big toe. The area is red and inflamed and tender. Gout is a disease in which a defect in uric acid metabolism causes an excess of the acid and its salts (sodium urate) to accumulate in the blood and ijoints. The big toe is commonly affected.*

the molecular structure of heredity. *See also* GENETIC CODE; GENETIC ENGINEERING

**genome** Entire complement of genetic material carried within the chromosomes of a single cell. In effect, a genome carries all the genetic information about an individual; it is coded in sequence by the DNA that makes up the chromosomes. The term has also been applied to the whole range of genes in a particular species. *See also* GENETICS

**geriatrics** Branch of medicine that deals with the problems of the elderly.

**germ** Popular term for any infectious agent. Germs can be bacteria, fungi, or viruses. In biology, it denotes a rudimentary stage in plant growth.

**German measles** (rubella) Viral disease usually contracted in childhood. Symptoms include a sore throat, slight fever, and pinkish rash. Women developing rubella during the first three months of pregnancy risk damage to the fetus. Immunization is recommended for girls who have not had the disease.

**giardiasis** Infection of the small intestine caused by the microscopic single-celled protozoan parasite *Giardia lamblia*. It is most commonly caused by eating contaminated food, drinking contaminated water, or by sexual contact. Symptoms include acute attacks of diarrhea and cramping abdominal pain.

**glaucoma** Incurable condition in which pressure within the eye is increased due to an excess of aqueous humor, the fluid within the chamber. It occurs when normal drainage of fluid is interrupted, posing a threat to vision. Most frequently found in

which proteins to make and in what quantities, the genetic code directly controls production of structural materials. It also codes for enzymes, which regulate all the chemical reactions in the cell, thus indirectly coding for the production of other cell materials as well. *See also* GENOME

**genetic engineering** Construction of a DNA molecule containing a desired gene. The gene is then introduced into a bacterial, fungal, plant, or mammalian cell, so that this cell produces the desired protein. It has been used to produce substances such as human growth hormone, insulin, and enzymes for biological washing powder. There are concerns about the ethics of genetic engineering both in human medicine and food technology. *See also* BIOTECHNOLOGY; CLONE

**genetics** Study of heredity. Geneticists study how the characteristics of an organism depend on its genes, how these characteristics pass down to the next generation, and how changes may occur through mutation. A person's behavior, learning ability, and physiology may be explained partly by genetics, although the environment has a considerable influence too. Gregor Mendel established the basic laws of inheritance. The discovery (1953) of DNA by Francis Crick, James Watson, and Maurice Wilkins revealed

people over the age of 40, glaucoma can be managed with drugs and surgery.

**glomerulonephritis** Group of kidney disorders featuring damage to the glomeruli (clusters of blood vessels in the kidney nephrons). Chronic forms may progress to kidney failure.

**gonorrhea** SEXUALLY TRANSMITTED DISEASE (STD) caused by the bacterium *Neisseria gonorrheae*, giving rise to inflammation of the genital tract. Symptoms include pain on urination and the passing of pus. Some infected women experience no symptoms. The condition is treated with antibiotics. If not treated, it may spread, causing sterility and ultimately threatening other organs in the body.

**gout** Form of ARTHRITIS, featuring an excess of uric acid crystals in the tissues. More common in men, it causes attacks of pain and inflammation in the joints, most often those of the feet or hands. It is treated with anti-inflammatories.

**gynecology** Area of medicine concerned with the female reproductive organs. Its study and practice is often paired with OBSTETRICS.

**hallucination** Apparent perception of something that is not actually present. Although they may occur in any of the five senses, auditory hallucinations and visual hallucinations are the most common. While they are usually symptomatic of psychotic disorders, hallucinations may also result from fatigue or emotional upsets and can sometimes be a side effect of certain drugs.

**hallucinogen** Drug that causes HALLUCINATIONS. Hallucinogens, such as mescaline, were used in primitive religious ceremonies. Today, drugs such as LSD are taken illicitly.

**haploid** Term used to describe a cell that has only one member of each CHROMOSOME pair. All human cells except GAMETES are diploid, having 46 chromosomes. Gametes are haploid, having 23 chromosomes. The body cells of many lower organisms, including algae and single-celled organisms, are haploid. *See also* MEIOSIS

**harelip** Congenital cleft in the upper lip caused by the failure of the two parts of the palate to fuse together. It is a congenital condition, often associated with CLEFT PALATE.

**H**

**hay fever** Seasonal ALLERGY induced by pollens. Symptoms include ASTHMA, itching of the nose and eyes, and sneezing. Symptoms are controlled with an ANTIHISTAMINE. *See* artwork, page 144

**headache** Pain felt in the skull. Most frequently caused by stress or tension, it may also signal other diseases, especially if associated with fever. *See also* MIGRAINE

**heart attack** (myocardial infarction) Death of part of the heart muscle due to the blockage of a coronary artery by a blood clot (thrombosis). It is accompanied by chest pain, sweating, and vomiting. Modern drugs treat abnormal heart rhythms and dissolve clots in the coronary arteries. **Heart failure** occurs when the heart is unable to pump blood at the rate necessary to supply body tissues and may be due to high BLOOD PRESSURE or CORONARY HEART DISEASE. Symptoms include shortness of breath, EDEMA, and fatigue. Treatment is with a DIURETIC and heart drugs. *See also* ANGINA; ARTERIOSCLEROSIS

**heart-lung machine** Apparatus used during some surgery to take over the function of the heart and lungs. It consists of a pump to circulate blood around the body and special equipment to add oxygen to the blood and remove carbon dioxide.

**heatstroke** Condition in which the body temperature rises above 106°F (41°C). It is brought on by extreme heat. In mild cases there may be lassitude and fainting; in severe cases, coma and death may ensue.

**Heimlich maneuver** FIRST AID technique, developed by Dr. Henry J. Heimlich for relieving blockage in the windpipe. The rescuer uses his or her arms to encircle the choking person's chest from behind, positioning one fist in the space just beneath the breastbone and covering it with the other hand. The rescuer then thumps the fist into the person's midriff.

**hemoglobin** Red-colored protein present in the ERYTHROCYTES (red-blood cells) of vertebrates. It carries oxygen to all cells in the body by combining with it to form oxyhemoglobin. Oxygen attaches to the heme part of the protein, which contains iron; the globin part is a globular protein.

**hemophilia** Hereditary BLOOD clotting disorder causing prolonged external or internal bleeding, often without apparent cause. **Hemophilia A** is caused by inability to synthesize blood factor VIII, a substance essential to clotting. This can be managed with injections of factor VIII. The rarer **hemophilia B** is caused by a deficiency of blood factor IX. The gene for both types is passed on almost exclusively from mother to son. Hemophilia A occurs in about 1 baby in 5,000 live births. Hemophilia B is occurs in about 1 in 30,000.

**hemorrhage** Loss of BLOOD from a damaged vessel. It may be external, flowing from a wound, or internal, as from internal injury or a bleeding ulcer. Blood loss from an artery is most serious, causing shock and death if untreated. Chronic bleeding can lead to ANEMIA. Internal bleeding is signaled by the appearance of blood in the urine or sputum.

**hemostasis** Process by which bleeding stops. BLOOD vessels constrict, PLATELETS

**H**

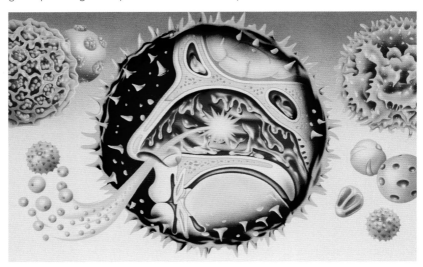

▲ *Hay fever Artwork depicting hay fever showing pollen grains (yellow/brown) entering the nasal cavity (red, center) of a sufferer. Pollen grains are released in their billions during plants' reproduction. Some people are sensitized to pollen in the air, and have an allergic reaction (rhinitis, or pollinosis) whenever they are in contact with it. Symptons are caused by a release of the chemical histamine in the body.*

► *Hodgkin's disease*
*Colored SEM of dividing Hodgkin's cells taken from the pleural effusions of a 55 year-old, male patient with 'mixed cellularity Hodgkin's disease'. The cells are grown in tissue culture. Hodgkin's disease is a cancer of the lymphoreticular system – the mediator of nonspecific cell defense mechanisms and the immune response. The various cell lines of the system are all subject to cancerous changes. In Hodgkin's disease, the nature of the cell line is unknown, but in the most common form the cell population is mixed. Magnification: ×572.*

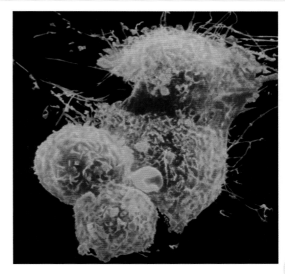

**H**

aggregate, and PLASMA coagulates to form filaments of FIBRIN.

**hepatitis** Inflammation of the liver, usually due to a generalized infection. Early symptoms include lethargy, nausea, fever, and muscle and joint pains. Five hepatitis viruses are known: A, B, C, D, and E. The most common single cause is the **hepatitis A** virus (HAV). More serious is infection with the **hepatitis B** virus (HBV), which can lead to chronic inflammation or complete failure of the liver and, in some cases, to liver cancer.

**hernia** Protrusion of an organ, or part of an organ, through its enclosing wall or connective tissue. Common hernias are a protrusion of an intestinal loop through the umbilicus (umbilical hernia), or protrusion of part of the stomach or esophagus into the chest cavity (hiatus hernia).

**heroin** Drug derived from MORPHINE. It produces similar effects to morphine, but acts more quickly and is effective in smaller doses. It is prescribed to relieve pain in terminal illness and severe injuries. Widely used illegally, it is more addictive than morphine and can lead eventually to death. *See also* DRUG ADDICTION

**herpes** Infectious disease caused by one of the herpes viruses. **Herpes simplex 1** infects the skin and causes cold sores. **Herpes zoster** attacks nerve ganglia, causing SHINGLES. The same virus is responsible for CHICKENPOX.

**histamine** Substance derived from the amino acid histidine, occurring naturally in many plants and in animal tissues, and released on tissue injury. It is implicated in allergic reactions that can be treated with ANTIHISTAMINE drugs.

**hives** (urticaria or nettle rash) Transient, itchy reddish or pale raised skin patches. Hives may be caused by an ALLERGY, by irritants such as sunlight, or by stress.

**Hodgkin's disease** (Hodgkin's lymphoma) Rare type of CANCER causing painless enlargement of the lymph glands, lymphatic tissue, and spleen, with subsequent spread to other areas. The cancer targets the B cells and the T cells of lymph tissue. Named for English pathologist Thomas Hodgkin (1798–1866), the treatment of Hodgkin's disease typically consists of RADIOTHERAPY, surgery, drug therapy, or a combination of these. It is curable if caught early.

**H**

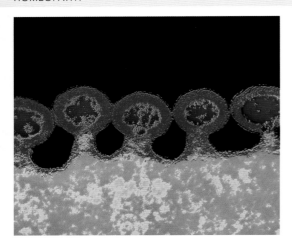

◀ *Human immunodeficiency virus (HIV)* Colored TEM of HIV viruses budding from a host T-lymphocyte white blood cell. The viruses are here acquiring their viral membrane (red) from the host cell membrane. HIV viruses infect T cells by injecting viral genetic material into the cell, causing it to form new HIV viruses. These new viruses bud out, damaging the cell membrane. The T cell may thus die, weakening the body's immune system. Magnification: ×72,000.

**homeopathy** Unorthodox medical treatment that involves administering minute doses of a drug or remedy which causes effects or symptoms similar to those that are being treated. German physician Christian Hahnemann popularized homeopathy in the 18th century.

**homeostasis** In biology, processes that maintain constant conditions within a cell or organism in response to either internal or external changes.

**hormone** Chemical substance secreted by living cells. Hormones affect the metabolic activities of cells in other parts of the body. In mammals, glands of the endocrine system secrete hormones directly into the bloodstream. Hormones exercise chemical control of physiological functions, regulating growth, development, sexual functioning, metabolism, and (in part) emotional balance. They maintain a delicate equilibrium that is vital to health. The hypothalamus, adjacent to the pituitary gland at the base of the brain, is responsible for overall coordination of the secretion of hormones. Most hormones are PROTEINS or STEROIDS. Hormones include THYROXINE, EPINEPHRINE, INSULIN, ESTROGEN, PROGESTERONE, and TESTOSTERONE. See also pages 56–57

**hormone replacement therapy (HRT)** Use of the female hormones PROGESTOGEN and ESTROGEN in women who are either menopausal or who have had both ovaries removed. HRT relieves symptoms of the MENOPAUSE; it also gives some protection against heart disease and OSTEOPOROSIS. Estrogen causes a thickening of the lining of the uterus, which may increase risk of cancer of the endometrium. Progestogen causes a regular shedding of the lining, similar to menstruation, which may lessen this risk.

**human immunodeficiency virus (HIV)** Organism that causes ACQUIRED IMMUNE DEFICIENCY SYNDROME (AIDS). A RETROVIRUS identified in 1983, HIV attacks the IMMUNE SYSTEM, leaving the person unable to fight off infection. There are two distinct viruses: **HIV-1**, which has now spread worldwide, and **HIV-2**, which is concentrated almost entirely in West Africa. Both cause AIDS. There are three main means of transmission: from person to person by sexual contact; from mother to baby during birth; and by contact with contaminated blood or blood products (for instance, during transfusions or when drug users share needles). People can carry the virus for many years before developing symptoms.

**Huntington's disease** (Huntington's chorea) Acute degenerative disorder. It is genetically transmitted and generally occurs in early middle age. It is caused by the presence of abnormally large amounts of glutamate and aspartate. Physical symptoms include loss of motor coordination. Mental deterioration can take various forms.

**hydrocephalus** Increased volume of cerebrospinal fluid (CSF) in the brain. A condition that exerts dangerous pressure on brain tissue, it can be due to obstruction or a failure of natural reabsorption. In babies, it is congenital; in adults, it may arise from injury or disease. It is treated by insertion of a shunting system to drain the CSF into the abdominal cavity.

**hydrotherapy** Use of water within the body or on its surface as a treatment of disease. It is often used in conjunction with PHYSIOTHERAPY.

**hyperglycemia** Condition in which blood-sugar level is abnormally high. It can occur in a number of diseases, most notably DIABETES. See also HYPOGLYCEMIA

**hypersensitivity** Condition in which a person reacts excessively to a stimulus. Most hypersensitive reactions are synonymous with an ALLERGY, the commonest being HAY FEVER.

**hypertension** Persistent high BLOOD PRESSURE. It can damage blood vessels and may increase the risk of strokes or heart disease. See also HYPOTENSION

**hyperthermia** Abnormally high body temperature, usually defined as being 105.8°F (41°C) or more. It is usually due to overheating (as in HEATSTROKE) or FEVER.

**hyperthyroidism** Excessive production of thyroid hormone, with enlargement of the thyroid gland (in the neck). Symptoms include protrusion of the eyeballs, rapid heart rate, high blood pressure, accelerated metabolism, and weight loss. See also HYPOTHYROIDISM; Endocrine system, pages 56–57

**hyperventilation** Rapid breathing that is not brought about by physical exertion. It reduces the carbon dioxide level in the blood, producing dizziness, tingling and tightness in the chest; it may cause loss of consciousness.

**hypochondria** (hypochondriasis) Neurotic condition characterized by an exaggerated concern with ill health. Hypochondriacs imagine they have serious diseases and often consult several doctors in the hope of a 'cure.'

**hypodermic syringe** Surgical instrument for injecting fluids beneath the skin into a muscle or blood vessel. It comprises a graduated tube containing a piston plunger, connected to a hollow needle. See also INJECTION

**hypoglycemia** Abnormally low level of sugar in the blood. It may result from fasting, excess INSULIN in the blood, or various metabolic and glandular diseases, notably DIABETES. Symptoms include dizziness, headache, sweating, and mental confusion. See also HYPERGLYCEMIA

**H**

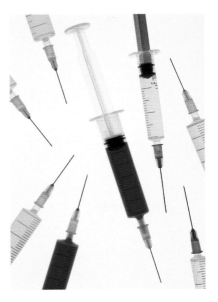

▲ *Hypodermic syringe* Photo showing a collection of syringes and hypodermic needles. Hypodermic syringes are used for administering vital fluids to patients.

**hypotension** Condition in which BLOOD PRESSURE is abnormally low. It is commonly seen after heavy blood loss or excessive fluid loss due to prolonged vomiting or diarrhea. It also occurs in many serious illnesses. Temporary hypotension may cause sweating, dizziness, and fainting. *See also* HYPERTENSION

**hypothermia** Fall in body temperature to below 95°F (35°C). Most at risk are newborns and the elderly. Insidious in onset, it can progress to coma and death. Hypothermia is sometimes induced during surgery to lower the body's oxygen demand. It occurs naturally in animals during hibernation.

**hypothyroidism** Deficient functioning of the thyroid gland. Congenital hypothyroidism can lead to cretinism in children. In adults, the condition is called **myxedema**. More common in women, it causes physical and mental slowness, weight gain, sensitivity to cold, and susceptibility to infection. It can be due to a defect of the gland or a lack of iodine in the diet. It is treated with the hormone thyroxine.

**hysterectomy** Removal of the uterus, possibly with surrounding structures. It is performed to treat fibroids or cancer or to put an end to heavy menstrual bleeding.

**immune system** System by which the body defends itself against disease. It involves many kinds of LEUKOCYTES (white blood cells) in the blood, lymph, and bone marrow. Some of the cells (B cells) make antibodies against invading microbes and other foreign bodies (ANTIGENS), or neutralize TOXINS produced by pathogens, while other antibodies encourage phagocytes andmacrophages to attack and digest invaders. T cells also provide a variety of functions in the immune system. *See also* INTERFERON; Immune system, pages 72–73

**immunity** In medicine, protection or resistance to DISEASE. Genetic factors and general health influence **innate** immunity. **Acquired** immunity is the body's second line of defense. An infecting agent stimulates the IMMUNE SYSTEM to respond to the presence of ANTIGENS. In **cell-mediated** immunity sensitized cells react directly with the antigen. This form of immunity is suppressed by HUMAN IMMUNODEFICIENCY VIRUS (HIV). Immunity may be induced artificially by IMMUNIZATION.

**immunization** Practice of conferring IMMUNITY against disease by artificial means. Passive immunity may be conferred by the injection of an antiserum containing antibodies. Active immunity involves vaccination with dead or attenuated (weakened) organisms to stimulate production of specific antibodies and so provide lasting immunity. Immunization against certain disease usually begins early in infanthood.

**immunoglobulin** Protein found in the BLOOD that plays a role in the IMMUNE SYSTEM. Immunoglobulins act as antibodies for specific ANTIGENS. They can be obtained from donor plasma and injected into people at risk of particular diseases.

**immunology** Study of IMMUNITY and ALLERGY. It is concerned with the prevention of disease by vaccination (**active** immunity) or by injections of antibodies (**passive** immunity). Allergic reactions result from an overactive response to harmless foreign substances such as dust, rather than to infective organisms.

**immunosuppressive drug** Any drug that suppresses the body's immune responses to infection or 'foreign' tissue. Such drugs are used to prevent rejection of trans-

▶ *Influenza Colored TEM of stages of cell infection of an influenza virus. The virus appears rounded, with a core of RNA. It has a spiked outer coat which allows the virus to attach to host cells. In the top frames, the virus attaches to the cell. In the lower frames, the virus penetrates the cell, infecting it and causing more influenza viruses to be produced. Magnification: ×50,000.*

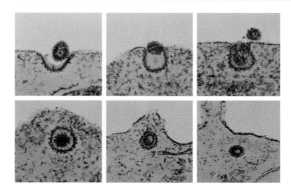

planted organs and to treat autoimmune disease and some cancers. STEROIDS, such as CORTISONE, which suppress lymphocyte cells, have been used in transplants.

**impetigo** Contagious skin condition caused by *staphylococcus aureus* infection. It causes multiple, spreading lesions with yellowish-brown crusts. The bacterium often lives harmlessly inside the nose.

**impotence** In men, the inability to perform SEXUAL INTERCOURSE through lack of an erection. It may be temporary or permanent, brought about by illness, injury, the effects of certain drugs, fatigue, or psychological factors.

**incubation period** In medicine, time-lag between becoming infected with a disease and the appearance of the first symptoms. In many infectious diseases, the incubation period is quite short – anything from a few hours to a few days – although it may also be very variable.

**indigestion** *See* DYSPEPSIA

**induction** Initiation of LABOR before it starts naturally. It involves perforating the fetal membranes and administering the hormone oxytocin to stimulate contractions of the uterus. It is undertaken where there is medical risk to the mother or baby in waiting for labor to begin naturally.

**infarction** Death of part of an organ caused by a sudden obstruction in an artery sup-

plying it. In a myocardial infarction (HEART ATTACK), a section of heart muscle dies.

**infection** Invasion of the body by disease-causing organisms that become established, multiply, and give rise to symptoms.

**infertility** Inability to reproduce. In a woman it may be due to a failure to ovulate (the release of an ovum for fertilization), obstruction of the Fallopian tube, or disease of the ENDOMETRIUM; in a man it is due to inadequate sperm production. *See also* pages 102–103

**inflammation** Reaction of body tissue to infection or injury, with resulting pain, heat, swelling, and redness. It occurs when damaged cells release HISTAMINE, which causes blood vessels at the damaged site to dilate. LEUKOCYTES invade the area to engulf bacteria. MACROPHAGES remove dead tissue, sometimes with the formation of pus.

**influenza** Viral infection mainly affecting the airways, with chesty symptoms, headache, joint pains, fever, and general malaise. It is treated by bed-rest and pain-killers. Vaccines are available to confer immunity to some strains.

**injection** In medicine, use of a syringe and needle to introduce drugs or other fluids into the body to diagnose, treat, or prevent disease. Most injections are either **intravenous** (into a vein), **intramuscular** (into a muscle), or **intradermal** (into the skin).

I

149

**insemination, artificial** Introduction of donor semen into a female's reproductive tract to bring about fertilization. First developed for livestock breeding, it is now routinely used to help infertile couples. *See also* IN VITRO FERTILIZATION (IVF)

**insomnia** Inability to sleep. It may be caused by anxiety, pain, or stimulants such as drugs.

**insulin** Hormone secreted by the islets of Langerhans in the pancreas and concerned with the control of blood-glucose levels. Insulin lowers the blood-glucose level by helping the uptake of glucose into cells, and by causing the liver to convert glucose to glycogen. In the absence of insulin, glucose accumulates in the blood and urine, resulting in DIABETES. In 1921 Frederick Banting and Charles Best isolated insulin. Its structure was discovered by Frederick Sanger. *See also* pages 56–57 and pages 88–89

**interferon** Protein produced by cells when infected with a VIRUS. Interferons can help uninfected cells to resist infection by the virus, and may also impede virus replication and protein synthesis. In some circumstances, they can inhibit cell growth. Human interferon is now produced by GENETIC ENGINEERING to treat some CANCERS, and HEPATITIS and MULTIPLE SCLEROSIS (MS).

**intoxication** Condition arising when the body is poisoned by any toxic substance, whether liquid, solid, or gas. Symptoms vary according to the ingested substance. The term is most commonly used in connection with alcohol consumption.

**intravenous drip** Apparatus for delivering drugs, blood and blood products, nutrients, and other fluids directly into the bloodstream. A hollow needle is inserted into an appropriate vein and then attached to tubing leading from a bag containing the solution.

**in vitro fertilization (IVF)** Use of artificial techniques to join an ovum with sperm outside (*in vitro*) a woman's body to help infertile couples to have children of their own. The basic technique of IVF involves removing ova from a woman's ovaries, fertilizing them in the laboratory, and then inserting them into her uterus. In **zygote intrafallopian transfer (ZIFT)**, a fertilized egg (zygote) is returned to the Fallopian tube, from which it makes its own way to the uterus. The zygote then divides to form an embryo. In **gamete intrafallopian transfer (GIFT)**, the ova are removed, mixed with sperm, then both ova and sperm are inserted into a Fallopian tube to be fertilized in the natural setting. The first 'test-tube baby', Mary Louise Brown, was born in England in

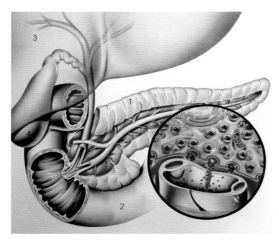

◀ *Insulin Artwork of the human pancreas showing production of the hormone insulin. The pancreas (1) is a tapered gland situated in the loop of the duodenum (2) and below the liver (3) and gall bladder (4). The pancreas has both digestive and hormonal functions. Its exocrine tissue produces digestive enzymes which pass into the intestine through the pancreatic duct (5). Production of the hormone insulin by the endocrine pancreas is shown inset. A beta cell (**red**) contained within an islet of Langerhans secretes insulin (**blue**) into a capillary (gray).*

▶ *In vitro fertilization* This technique was developed to help women who are unable to conceive. The mother, or egg donor, is given hormone treatment to cause many ovarian follicles to develop (*1*). While the ovary is observed through a laparoscope, a needle (*2*) is inserted into each follicle, which is then transferred (*3*) to a petri dish where sperm are added (*4*). As the zygote divides (*5*) to 16 cells, it is introduced into the uterus through a plastic tube so that it can implant in the wall (*6*) and progress normally.

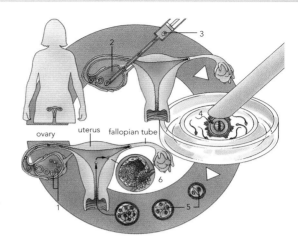

ovary  uterus  fallopian tube

1978. *See also* INFERTILITY; Fertilization and Implantation, pages 102–103

**IVF** Abbreviation of IN VITRO FERTILIZATION

effects on physiological and psychological rhythms. Symptoms include fatigue, confusion, mood alterations, irritability, sleep disturbances, and other signs of stress.

J

K

**jaundice** Yellowing of the skin and the whites of the eyes, caused by an excess of BILE pigment in the blood. Mild jaundice is common in newborn babies. In adults, jaundice may occur when the flow of bile from the liver to the intestine is blocked by an obstruction such as a GALL-STONE, or in diseases such as CIRRHOSIS, HEPATITIS, or ANEMIA.

**jet lag** Phenomenon experienced when the body clock is disrupted by the sudden change in time zones, causing adverse

**keratin** Fibrous protein present in large amounts in skin cells, where it serves as a protective layer. Hair and fingernails are made up of cells filled with keratin.

**kidney machine** (artificial kidney) Equipment designed to remove toxic wastes from the blood in kidney failure. Plastic tubing is used to pipe blood from the body into the machine, where waste products are filtered out by DIALYSIS. In 1943 the first kidney machine was built by Dutch physician Willem Kolff.

**labor** In childbirth, stage in the delivery of the fetus at the end of PREGNANCY. In the first stage, contractions of the uterus begin and the cervix dilates in readiness; the sac containing the amniotic fluid ruptures. In the second stage, the contractions strengthen and the baby is propelled through the birth canal. The third stage is the expulsion of the placenta and fetal membranes, together known as the afterbirth. *See also* Childbirth, pages 106–107

**lactation** Secretion of milk to feed the young. In pregnant women, hormones induce the breasts to enlarge, and PROLACTIN stimulates breast cells to begin secreting milk. The milk appears in the breast immediately after the birth of the baby. The hormone oxytocin controls the propulsion of milk from the breast.

**lactose** (milk sugar) Disaccharide present in milk, made up of a molecule of glucose linked to a molecule of galactose.

**lanolin** Purified, fatlike substance derived from sheep's wool and used with water as a base for ointments and cosmetics.

**laryngitis** Inflammation of the larynx and vocal cords. Symptoms include a sore throat, hoarseness, coughing, and breathing difficulties. It is usually due to a respiratory tract infection but may also be caused by exposure to irritant gases or chemicals. *See also* Larynx and Speech, pages 80–81

**laser surgery** Surgery carried out using a laser. The high energy in an narrow laser beam can burn through body tissues to make a fine 'cut.' The heat also seals blood vessels, so there is less bleeding than when a knife is used. Less powerful lasers can remove marks from the skin. Some forms of skin cancer are treated in this way.

**Lassa fever** Acute viral disease, classified as a hemorrhagic fever. The virus, first detected in 1969, is spread by a species of rat found only in West Africa. It is characterized by internal bleeding, with fever, headache, and muscle pain.

◀ *Lactose Polarized light micrograph of crystals of the sugar lactose found in milk. Lactose is a white sugar occurring in solution in the milk of all animals. Human milk contains 6% lactose; cow's milk contains 4%. It is commercially prepared by evaporation of the whey obtained in cheesemaking. Like table sugar (sucrose), it is a disaccharide consisting of two basic sugar units: glucose and galactose. Magnification: ×63.*

L

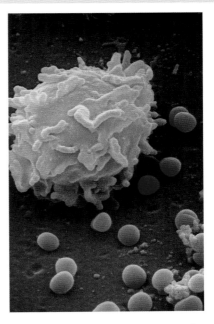

▶ **Leukocyte** *Colored SEM of a leukocyte (white blood cell) surrounded by* Staphylococcus aureus *bacteria. The leukocyte (mauve) is part of the body's immune response against the presence of the bacteria (red). Magnification: ×5,500.*

**laxative** Any agent used to counteract constipation. There are various kinds available, including bulk-forming drugs, stimulant laxatives, fecal softeners, and saline purgatives.

**L-dopa** (levodopa) Naturally occurring amino acid used to relieve some symptoms of PARKINSON'S DISEASE. It sometimes suppresses the trembling, unsteadiness, and slowness of movement that characterize the condition.

**legionnaire's disease** Lung disease caused by infection with the bacterium *Legionella pneumophila*. It takes its name from the serious outbreak that occurred during a convention of the American Legion held in Philadelphia, PA, in 1976. The bacterium thrives in water and may be found in defective heating, ventilation, and air-conditioning systems. It is inhaled in fine water droplets present in the air.

**Leishmaniasis** Insect-borne disease carrying a high mortality and caused by infection with the parasite *Leishmania donovani*, transmitted by sand flies. The spleen is particularly affected and becomes enlarged. Symptoms include fever, anemia, and wasting. The disease occurs primarily in the Mediterranean region, Africa, Asia, and Central and South America.

**leprosy** (Hansen's disease) Chronic, progressive condition affecting the skin and nerves, caused by infection with the microorganism *Mycobacterium leprae*. There are two main forms. **Lepromatous** leprosy is a contagious form in which raised nodules appear on the skin and there is thickening of the skin and peripheral nerves. In **tuberculoid** leprosy, there is loss of sensation in parts of the skin, sometimes with loss of pigmentation and hair. Now confined almost entirely to the tropics, leprosy can be treated with a combination of drugs, but the nerve damage is irreversible.

**leukemia** Any of a group of CANCERS in which the bone marrow and other blood-forming tissues produce abnormal numbers of immature or defective LEUKOCYTES. This overproduction suppresses output of normal blood cells and platelets, leaving the person vulnerable to infection, anemia, and bleeding. **Acute lymphoblastic leukemia** (ALL) is predominantly a disease of childhood; **acute myelogenous leukemia** (AML) is mainly seen in older adults. Both forms are potentially curable. *See illustration*, page 154

**leukocyte** (white BLOOD cell) Colorless structure containing a nucleus and cytoplasm. There are two types of leukocytes – LYMPHOCYTES and PHAGOCYTES. Normal blood contains 5,000–10,000 leukocytes per cu mm. Excessive numbers of leukocytes are seen in such diseases as LEUKEMIA. *See also* ANTIBODY; IMMUNE SYSTEM

**L**

153

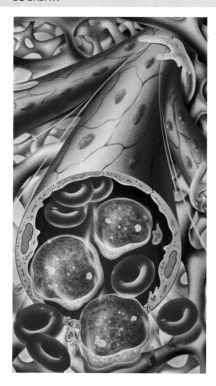

◀ *Leukemia Illustration of chronic myeloid leukemia. Seen passing between red blood cells in the interior of this branching blood vessel are the myeloblast cells that establish this type of blood cancer. Myeloblasts (pink) are immature white blood cells that proliferate uncontrollably. Two phases normally occur: a chronic phase that may last some years; and a more malignant, acute phase.*

and nose. The systemic or disseminated form varies in severity. Nine times more common in women than in men, the disease is treated mainly with corticosteroids.

**lupus vulgaris** Tuberculous infection of the skin. Often starting in childhood, it is characterized by the formation of brownish nodules, leading to ulceration and extensive scarring. Treatment is with antituberculous drugs.

**lyme disease** Condition caused by a spirochete (spiral-shaped bacterium) transmitted by the bite of a tick that lives on deer. It usually begins with a red rash, often accompanied by fever, headache, and pain in the muscles and joints. Untreated, the disease can lead to chronic arthritis, and there may also be involvement of the nervous system, heart, liver, or kidneys. It is treated with ANTIBIOTICS.

**lymph gland** (lymph node) Mass of tissue occurring along the major vessels of the lymphatic system. They are filters and reservoirs that collect harmful material, notably bacteria and other disease organisms, and often become swollen when the body is infected. *See also* Lymphatic system, pages 70–71

**lymphocyte** Type of LEUKOCYTE (white blood cell) found in vertebrates. Produced in bone marrow, they are mostly found in the lymph nodes and blood and around infected sites. In human beings, lymphocytes make up *c.*25% of white blood cells and play an important role in combating disease. The **B-lymphocytes** produce antibodies and the **T-lymphocytes** maintain IMMUNITY. *See also* ANTIBODY

**lockjaw** *See* TETANUS

**long sightedness** (hyperopia) Defect of vision that causes distant objects to be seen more clearly than nearby ones. In a long sighted person, the focusing distance of the eyeball is too short and, as a result, light rays entering the eye strike the retina before they can be properly focused. Long sightedness is corrected by convex lenses. *See also* MYOPIA

**lumbago** Pain in the lower (lumbar) region of the back. It is usually due to strain or poor posture. When associated with SCIATICA, it may be due to a slipped disk. *See also* RHEUMATISM

**lupus erythematosus** Autoimmune disease affecting the skin and connective tissue. The discoid form causes red patches covered with scales, often on the cheeks

and also used as an antacid, is often also called magnesia.

**magnetic resonance imaging (MRI)** Diagnostic scanning system, based on the use of powerful magnets, which produces images of soft tissues in the body. Magnetic resonance imaging (MRI) is invaluable for producing images of the brain and spinal cord in particular. The scanner's magnet causes the nuclei within the atoms of the patient's body to line themselves up in one direction. A brief radio pulse is then beamed at the nuclei, causing them to spin. As they realign themselves to the magnet, they give off weak radio signals that can be recorded and converted electronically into images.

**macrophage** Large LEUKOCYTE (white blood cell) found mainly in the liver, spleen, and lymph nodes. It engulfs foreign particles and microorganisms by phagocytosis. Working together with other lymphocytes, it forms part of the body's defense system.

**mad cow disease** Popular name for BOVINE SPONGIFORM ENCEPHALOPATHY (BSE)

**magnesia** Magnesium oxide (MgO), a white, neutral, stable powder formed when magnesium burns in oxygen. It is used medicinally in stomach powders. **Magnesium carbonate**, found as magnesite

**malaria** Parasitic disease resulting from infection with one of four species of *Plasmodium* protozoan (unicellular organism). Transmitted by the *Anopheles* mosquito, it is characterized by fever and enlargement of the spleen. Attacks of fever, chills, and sweating typify the disease, and recur as new generations of parasites develop in the blood. The original antimalarial drug, QUININE, gave way to synthetics such as chloroquine. With 270 million people infected, malaria is one of

▶ *Malaria An infected mosquito injects thousands of* Plasmodium *organisms into the bloodstream when it bites a human (1). These penetrate liver cells, multiply and cause cell rupture (2). Organisms may re infect liver cells (3), but usually progress to infect red blood cells (4,5). Male and female parasites appear in the red blood cells (6). Another mosquito bites the human and takes infected blood (7). Fertilization occurs within the mosquito, the 'embryo' penetrating the stomach wall (8). The cycle continues (9).*

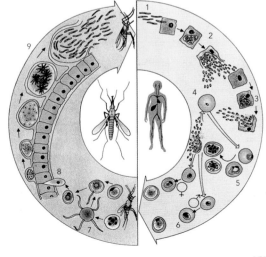

the most widespread diseases, claiming two million lives a year.

**malnutrition** Condition resulting from a diet that is deficient in components such as proteins, fats, or carbohydrates. It can lead to deficiency diseases, increased vulnerability to infection, and death. *See also* FAMINE

**manic depression** (bipolar disorder) Mental illness featuring recurrent bouts of depression, possibly alternating with periods of mania. Depressive and manic symptoms may alternate in a cyclical pattern, be mixed, or be separated by periods of remission and disturbances of thought and judgment.

**mastectomy** In surgery, removal of all or part of the female breast. It is performed to treat CANCER. **Simple** mastectomy involves the breast alone. If the cancer has spread, **radical** mastectomy may be undertaken, removing also the lymphatic tissue from the armpit.

**ME** (myalgic encephalomyelitis) Condition defined as extreme fatigue that persists for six months or more. Also known as **chronic fatigue syndrome** or **postviral fatigue syndrome**, it may be accompanied by other nonspecific symptoms, such as fever, headache, muscle and joint pains, dizziness, and mood changes. Often occurring after a VIRUS infection, it ranges in severity from chronic weariness to total physical collapse. The cause of ME is unknown.

**measles** (rubeola) Extremely infectious viral disease of children. The symptoms (fever, catarrh, skin rash, and spots inside the mouth) appear about two weeks after exposure. Hypersensitivity to light is characteristic. Complications such as pneumonia occasionally occur, and middle-ear infection is also a hazard. Vaccination produces life long IMMUNITY.

**medicine** Practice of the prevention, diagnosis, and treatment of disease or injury; the term also applies to any agent used in the treatment of disease. Medicine has been practiced since ancient times, but the dawn of modern Western medicine coincided with accurate anatomical and physiological observations first made in the 17th century. By the 19th century, practical diagnostic procedures had been developed for many diseases; BACTERIA had been discovered and research undertaken for the production of immunizing serums in attempts to eradicate disease. The great developments of the 20th century included the discovery of PENICILLIN and INSULIN, CHEMOTHERAPY (the treatment of various diseases with specific chemical agents), new surgical procedures including organ transplants, and sophisticated diagnostic

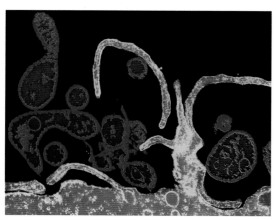

◄ **Measles** Colored TEM of measles viruses budding off the surface of an infected cell (bottom, pink/yellow). The lipoprotein envelope the viruses (red), surrounding the nucleocapsid (blue). The envelopes are acquired from the host cell's cytoplasmic membrane as the viruses bud from the surface. They enclose the nucleocapsid: a helical structure consisting of a single-stranded RNA core surrounded by protein. Measles is very infectious and mainly affects children. Magnification: ×14,400.

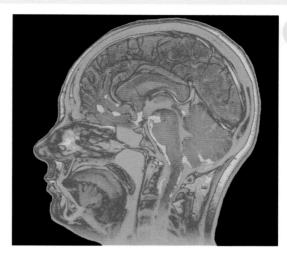

► *Meningitis* Colored MRI scan of a section through the head of a 34-year-old woman. The brain is seen, with the folded cerebrum (at top), brainstem (center), and cerebellum (center right). The spinal cord passes into the neck. Meningitis infection is colored yellow on surfaces of the brain and spinal cord. Meningitis is an inflammation of the meninges – the membranes that cover and protect the brain and spinal cord. It is caused by either bacterial or viral infection.

devices such as radioactive TRACERS and scanners. Alternative medicine, such as OSTEOPATHY, HOMEOPATHY, or ACUPUNCTURE, some of which are hundreds of years old, is becoming increasingly popular, and some therapies are being accepted within conventional medicine.

**meiosis** In biology, the process of cell division that reduces the CHROMOSOME number from DIPLOID to HAPLOID. The first division halves the chromosome number in the cells; the second then forms four haploid 'daughter' cells, each containing a unique configuration of the parent cells' chromosomes. In the majority of higher organisms, the resulting haploid cells are the GAMETES, ova and sperm. In this way, meiosis enables the genes from both parents to combine in a single cell without increasing the overall number of chromosomes. *See also* MITOSIS

**melanin** Dark pigment found in the skin, hair, and parts of the eye. The amount of melanin determines skin color. Absence of melanin results in an ALBINO.

**melanoma** Type of skin CANCER. They are highly malignant tumors formed by melanocytes, cells in the skin that make the dark pigment MELANIN. They may also occur in mucous membranes and in the eye. Untreated, they may spread to the liver and lymph nodes. Excessive exposure to sunlight has been cited as a causative factor. Treatment is usually by surgery.

**Ménière's disease** Chronic condition of the inner ear affecting hearing and balance. Symptoms include deafness, vertigo and ringing in the ears (tinnitus). Caused by an excessive amount of fluid in the inner ear, it usually occurs in middle age or later. It is generally treated with ANTIHISTAMINE drugs.

**meningitis** Inflammation of the **meninges** (membranes) covering the brain and spinal cord, resulting from infection by bacteria or viruses. **Bacterial meningitis** is less common but more serious than its viral counterpart. The germs live naturally at the back of the nose and throat of one in ten people and can be spread by close prolonged contact. Symptoms include headache, fever, nausea, and stiffness of the neck. The disease can vary from mild to lethal.

**menopause** Stage in a woman's life marking the end of her reproductive years, when the MENSTRUAL CYCLE becomes irregular and finally ceases, generally around the age of 50. It may be accompanied by side effects such as hot flushes, excessive bleeding, and emotional upset. Hormone

**M**

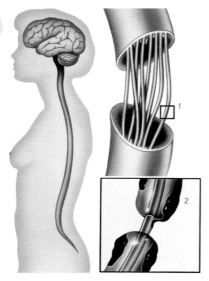

See also pages 100–103

◄ *Multiple sclerosis (MS) Illustration of the brain and spinal cord of a woman suffering from MS. Also seen is a magnified view of the spinal cord showing nerve fibers (1). At lower right is a magnified view of an axon within one of the nerve fibers covered with a protective myelin sheath. In MS, patches of the myelin sheath are destroyed. These patches of destroyed myelin are known as plaques (2). The damaged fibers affect nerve impulses.*

**metabolism** Chemical and physical processes and changes continuously occurring in an organism. These processes include the breakdown of organic matter (**catabolism**), resulting in energy release and the synthesis of organic components (**anabolism**) to store energy and build and repair tissues. *See also* RESPIRATION

**microbiology** Study of microorganisms, their structure, function and significance. Mainly concerned with VIRUSES, bacteria, protozoa, and fungi, it has wide applications in medicine and the food industry.

**microsurgery** Delicate surgery performed under a binocular microscope using specialized instruments, such as microneedles as small as 2 mm (0.08 in) long, ENDOSCOPES, and lasers. It is used in a number of specialized areas, including the repair of nerves and blood vessels; eye, ear, and brain surgery; and the re-attachment of severed parts.

**migraine** Recurrent attacks of throbbing headache, mostly on one side only, often accompanied by nausea, vomiting, and visual disturbances. It results from changes in diameter of the arteries serving the brain. More common in women, it is seen usually in young adults and may often run in families. Attacks, which may last anything from two to 72 hours, are often associated with trigger factors, such as certain foods (especially chocolate), missed meals, consumption of alcohol, fatigue, exposure to bright light, or use of the contraceptive pill. It can be treated, or in some cases prevented, with various drugs.

**mineral** In NUTRITION, naturally occuring substances that are an important dietary

replacement therapy (HRT) is designed to relieve menopausal symptoms.

**menstrual cycle** In humans and some higher primates of reproductive age, the stage during which the body prepares for pregnancy. In humans, the average cycle is 28 days. At the beginning of the cycle, hormones from the pituitary gland stimulate the growth of an ovum contained in a follicle in one of the two ovaries. At approximately midcycle, the follicle bursts, the ovum releases (**ovulation**) and travels down the FALLOPIAN TUBE to the uterus. The follicle (now called the **corpus luteum**) secretes two hormones, PROGESTERONE and ESTROGEN, during this secretory phase, and the ENDOMETRIUM thickens, ready to receive the fertilized ovum. If fertilization (conception) does not occur, the corpus luteum degenerates, hormone secretion ceases, the endometrium breaks down, and menstruation occurs in the form of a loss of blood. In the event of conception, the corpus luteum remains and maintains the endometrium with hormones until the placenta is formed. In humans, the onset of the menstrual cycle occurs at PUBERTY; it ceases with the MENOPAUSE (around 50 years of age). *See also* pages 100–103

**M**

element. The 'major' minerals are calcium and phosphorus as they are required in large amounts, particularly to aid growth of teeth and bones. '**Trace**' minerals are needed in tiny amounts, and include iron, sodium, chlorine, sulfur, zinc, copper, manganese, and magnesium. All minerals are consumed in a well-balanced diet. *See also* VITAMIN

**minimal access surgery** Operations that do not involve cutting open the body in the traditional way. Minimal access (**keyhole**) procedures are performed either by means of an ENDOSCOPE, or by passing miniature instruments through a fine catheter into a large blood vessel. The surgical laser is also used.

**miscarriage** Popular term for a spontaneous ABORTION, the loss of a fetus from the uterus

**mitosis** Cell division resulting in two genetically identical 'daughter' cells with the same number of chromosomes as the parent cell. Mitosis is the normal process of tissue growth, and is also involved in asexual reproduction. See also meiosis

**mongolism** *See* DOWN'S SYNDROME

**mononucleosis, infectious** (glandular fever) Acute disease, usually affecting young people, caused by the Epstein-Barr virus. Mononucleosis produces an increased number of white cells (monocytes) in the blood. Symptoms include fever, painful enlargement of the lymph nodes, and pronounced lassitude.

**morphine** White, crystalline alkaloid derived from OPIUM. It depresses the CENTRAL NERVOUS SYSTEM and is used as an ANALGESIC for severe pain. An addictive drug, its use is associated with a number of side effects, including nausea. Morphine was first isolated in 1806. *See also* HEROIN

**multiple sclerosis (MS)** Incurable disorder of unknown cause in which there is degeneration of the myelin sheath that surrounds nerves in the brain and spinal cord. Striking mostly young adults, symptoms may include unsteadiness, loss of coordination, and speech and visual disturbances. MS sufferers typically have relapses and remissions over many years. MS affects about one in 2,000 people.

**mumps** Viral disease, most common in children, characterized by fever, pain, and swelling of one or both parotid salivary glands (located just in front of the ears). The symptoms are more serious in adults, and in men inflammation of the testes (orchitis) may occur, with the risk of sterility.

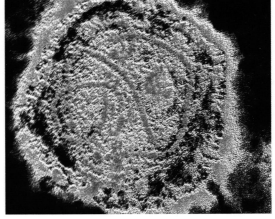

► *Mumps Colored TEM of a section through a mumps virus in a capsule. The long, thin chains (red) contain ribonucleic acid (RNA) genetic material. They are seen in a capsule (pink). The mumps virus is a paramyxovirus. The infection usually causes swelling of the parotid glands of the neck. It is a common childhood disease, and symptoms normally subside within three days.*

**M**

**muscular dystrophy** Any of a group of hereditary disorders in which the characteristic feature is progressive weakening and ATROPHY of the muscles. Muscle fibers degenerate, to be replaced by fatty tissue. The most common type, **Duchenne muscular dystrophy**, affects about one male infant in 3,000. There is a milder form called **Becker's muscular dystrophy**, which develops in young adults.

**myopia** (short sightedness) Common disorder of vision in which near objects are seen sharply, but distant objects are hazy. It is caused either by the eyeball being too long or the eye's lens being too powerful, so that light rays entering the eye focus in front of the retina. It is easily corrected with concave lenses in spectacles or contact lenses.

**myxedema** Disease caused by deficient function of the thyroid gland, resulting in fatigue, constipation, dry skin, a tendency toward weight-gain and, in the later stages, a dulling of mental lucidity. It mostly affects middle-aged women. Treatment involves the administration of the thyroid hormone thyroxine. *See also* pages 56–57

**narcotic** Any drug that induces sleep and/or relieves pain. The term is used especially in relation to OPIUM and its derivatives. These drugs have largely been replaced as sedatives because of their addictive properties, but they are still used for severe pain, notably in terminal illness. Other narcotics include alcohols and BARBITURATES.

**National Health Service (NHS)** In Britain, system of state provision of health care established in 1948. The NHS undertook to provide free, comprehensive coverage for most health services, including hospitals, general medical practice, and public health facilities. It is administered by the Department of Health. General practitioners (GPs) have registered patients; they may also have private patients and may contract out of the state scheme altogether. They refer patients, when necessary, to specialist consultants in hospitals. Health visitors, such as midwives and district nurses, are the third arm of the service. Hospitals are administered by regional boards. In 1990, the Conservative government introduced the concept of the 'internal market' into health care, establishing GP fund-holding practices and NHS Trusts independent of local health authority control. In 1997, the Labour government announced that it would replace the internal market with primary care groups consisting of GPs and community nurses. The NHS is the largest employer in the UK.

**naturopathy** System of medical therapy that relies exclusively on the use of natural treatments, such as exposure to sunlight, fresh air, and a healthy diet of organically grown foods. People practicing naturopathy prefer the use of herbal remedies to manufactured drugs.

**necrosis** Death of plant or animal tissue. It can be caused by disease, injury, or interference with the blood supply.

**nephritis** Inflammation of the kidney. It is a general term used to describe a condition rather than any specific disease. It can progress to kidney failure.

**nervous breakdown** Popular term for a mental and emotional crisis in which the person either is unable or feels unable to function normally. It is an imprecise term and refers to any of a range of conditions. Nervous breakdowns are usually caused by periods of extreme emotional stress.

**neuralgia** Intense pain from a damaged nerve, possibly tracking along its course.

▶ *Neurotransmitter Colored SEM showing the junction sites, also known as synapses, between nerve fibers and a nerve cell grown in culture. Two nerve fibers colored purple run diagonally. At top is the surface of a nerve cell (yellow) to which the nerve fibers have formed synapses. Impulses travel down these fibers into the nerve cell, typical of dendrites. The fibers branch before forming their junctions which appear as terminal swellings (top left). Across these swellings neurotransmitter chemicals pass into the cell in response to nerve fiber impulses and in this way the cell is stimulated. Magnification: ×10,000.*

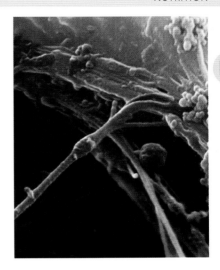

**N**

Forms include trigeminal neuralgia, which features attacks of stabbing pain in the mouth area, and postherpetic neuralgia following an attack of SHINGLES.

**neurology** Branch of medicine dealing with the diagnosis and treatment of diseases of the nervous system. *See also* Nervous system, pages 34–55

**neurosis** Emotional disorder such as anxiety, depression, or various phobias. It is a form of mental illness in which the main disorder is of mood, but the person does not lose contact with reality as happens in PSYCHOSIS.

**neurotransmitter** Any one of several dozen chemicals involved in communication between neurons or between a nerve and muscle cells. When an electrical impulse arrives at a nerve ending, a neurotransmitter is released to carry the signal across the synapse (specialized junction) between the nerve cell and its neighbor. Some drugs work by disrupting neurotransmission. *See also* pages 34–55

**nicotine** Poisonous alkaloid obtained from the leaves of tobacco, used in agriculture as a pesticide and in veterinary medicine to kill external parasites. Nicotine is the principal addictive agent in smoking tobacco. *See also* ADDICTION

**nitroglycerine** Oily liquid used in medicine (as nitroglycerin) to relieve the symptoms of ANGINA.

**norepinephrine** Hormone secreted by nerves in the autonomic nervous system and by the adrenal glands (on top of the kidneys). It slows the heart rate and constricts small arteries, thus raising the blood pressure. It is used therapeutically to combat the fall in blood pressure that accompanies shock.

**nursing** Profession that has as its general function the care of people who, through ill health, disability, immaturity, or advanced age, are unable to care for themselves. The Christian Church emphasized caring for the sick, and many religious orders performed such 'acts of mercy.' In the 19th century, Florence Nightingale revealed the need for reforms in nursing, and by the end of the century, England and the Unied States had adopted some of her principles. Modern nursing provides a broad range of services, with standards set by relevant professional bodies.

**nutrition** Processes by which plants and animals take in and make use of food substances. The science of nutrition involves identifying the kinds and amounts of nutrients necessary for growth and health. Nutrients are generally divided into PROTEINS, CARBOHYDRATES, FATS, MINERALS, and VITAMINS. *See also* DIET

**obesity** Condition of being overweight, generally defined as weighing 20% or more above the recommended norm for the person's sex, height, and build. People who are overweight are at increased risk of disease and have a shorter life-expectancy than those of normal weight.

**obstetrics** Branch of medicine that deals with pregnancy, childbirth, and the care of women following delivery.

**occupational therapy** Development of practical skills to assist patients recovering from illness or injury. Therapists oversee a variety of pursuits, from the activities of daily living (ADLs), such as washing and dressing, to hobbies and crafts.

**odontology** Study of the structure, development ,and diseases of the teeth. It is closely allied with DENTISTRY.

**onchocerciasis** (river blindness) Tropical disease of the skin and connective tissue, caused by infection with filarial worms; it may also affect the eyes, causing blindness. It is transmitted by blood-sucking blackflies found in Central and South America and Africa.

▶ **Opium** *Colored X-ray of the seed capsule of an opium poppy,* Papaver somniferum. *This is the structure from which opium is collected: the circumference of the capsule is slit, allowing a milky latex to ooze out. The latex hardens to form a plastic gummy substance on exposure to air, which is scraped off and formed into balls.*

**oncogene** Gene that, by inducing a cell to divide abnormally, contributes to the development of CANCER. Oncogenes arise from gene mutations (proto oncogenes), which are present in all normal cells and in some viruses.

**oncology** Specialty concerned with the diagnosis and treatment of CANCER.

**ophthalmology** Branch of medicine that specializes in the diagnosis and treatment of diseases of the eye.

**ophthalmoscope** Instrument for examining the interior of the eye, invented by Hermann von Helmholtz in 1851.

**opium** Drug derived from the unripe seed pods of the opium poppy. Its components and derivatives have been used as NARCOTICS and ANALGESICS for many centuries. It produces drowsiness and euphoria and reduces pain. MORPHINE and CODEINE are opium derivatives.

**optometry** Testing of vision in order to prescribe corrective eyewear, such as spectacles or contact lenses. It is distinct from OPHTHALMOLOGY.

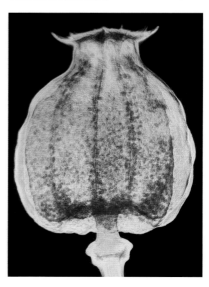

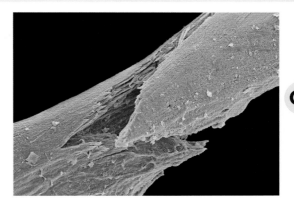

▶ *Osteoporosis* Colored SEM of fractured bone tissue from a patient suffering from osteoporosis (brittle bone disease). Osteoporosis causes a reduction in overall bone mass and an increase in the bone's porosity, making it more brittle and likely to fracture. It commonly affects the elderly and postmenopausal women, who experience a decrease in levels of the hormone estrogen. It may develop after injury or infection.

**orgasm** Physiological culmination of sexual stimulation, marked by general release of muscular tension and waves of contractions causing climactic spasms of vaginal muscles in the female and ejaculation (the release of SEMEN) in the male.

**orthodontics** See DENTISTRY

**orthopedics** Branch of medicine that deals with the diagnosis and treatment of diseases, disorders, and injuries of bones, muscles, tendons, and ligaments.

**osteoarthritis** Degenerative disease of the joints, causing pain on movement. It is the most common form of ARTHRITIS. Eighty percent of people aged 65 show evidence of the condition. Women tend to be more severely affected than men. The disease most commonly affects the spine, the knee joints and the hip joints. The basic change in the affected joints is the loss of the articular cartilage, which normally protects the ends of the bones and provides a smooth working surface for movement. Small spurs of bone, known as osteophytes, develop at the margins of the exposed ends of bones.

**osteomyelitis** Infection of the bone, sometimes spreading along the marrow cavity. Rare except in diabetics, it can arise from a compound fracture, where the bone breaks through the skin, or from infection elsewhere in the body. It is accompanied by fever, swelling, and pain. The condition may be treated with immobilization, ANTIBIOTICS, and surgical drainage.

**osteopathy** System of alternative medical treatment based on the use of physical manipulation of joints to rectify damage caused by mechanical stresses.

**osteoporosis** Condition where there is loss of bone substance, resulting in brittle bones. It is common in older people, especially in women after the MENOPAUSE. It may also occur as a side effect of prolonged treatment with corticosteroid drugs. There is no cure, but it may be treated with calcium supplements. HORMONE REPLACEMENT THERAPY (HRT) may help to prevent its occurence in postmenopausal women.

**otosclerosis** Inherited condition in which overgrowth of bone in the middle ear causes deafness. It is gradual in onset and twice as common in women as in men. Surgery can rebuild the sound conduction mechanism.

**oxygen debt** Insufficient supply of oxygen in the muscles following vigorous exercise. This reduces the breakdown of food molecules that generate energy, causing the muscles to overproduce lactic acid creating a sensation of fatigue and sometimes muscular cramp.

**oxytocin** Hormone produced by the pituitary gland in women during pregnancy. It

163

**P**

stimulates the muscles of the uterus, initiating the onset of labor and maintaining contractions during childbirth. It also stimulates lactation. *See also* pages 58–59 and pages 104–105

**pacemaker** (sino atrial node) Specialized group of cells in the vertebrate heart that contract spontaneously, setting the pace for the heartbeat itself. If it fails, it can be replaced by an artificial pacemaker – an electronic unit that stimulates the heart by means of tiny electrical impulses.

**pain** Unpleasant sensation signaling actual or threatened tissue damage as a result of illness or injury; it can be **acute** (severe but short-lived) or **chronic** (persisting for a long time). Pain is felt when

specific nerve endings are stimulated. Pain is treated in a number of ways, most commonly by drugs known as ANALGESICS.

**pap test** *See* CERVICAL SMEAR

**paralysis** Weakness or loss of muscle power; it can vary from a mild condition to complete loss of function and sensation in the affected part. It can be associated with almost any disorder of the nervous system, including brain or spinal cord injury, infection, stroke, poisoning, or progressive conditions such as a tumor or motor neurone disease. Paralysis is rarely total.

**paraplegia** PARALYSIS of both legs. It is usually due to spinal cord injury, and often accompanied by loss of sensation below the site of the damage.

**parasite** Organism that lives on or in another organism (the host) upon which it depends for its survival; this arrangement may be harmful to the host. A parasite that lives in the host is called an **endoparasite**. A parasite that survives on the host's exterior is known as an **ectoparasite**. In **parasitoidism**, the relationship results in the death of the host.

**Parkinson's disease** Degenerative brain disorder characterized by tremor, muscular rigidity, and poverty of movement and

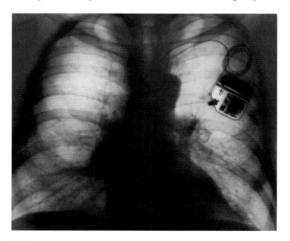

◀ *Pacemaker Colored chest X-ray showing a surgically implanted heart pacemaker (at upper right). The pacemaker supplies electrical impulses along a lead to the heart to keep it beating regularly. Pacemakers are fitted in patients with a malfunctioning sino atrial node, the part of the heart which initiates each heartbeat, or in patients with a heart block which impairs the nerve impulses generated by the node. Heart pacemakers may provide a regular impulse or discharge only when a heartbeat is missed.*

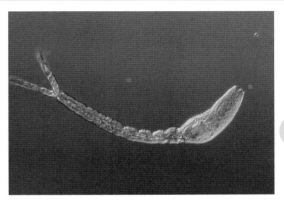

► *Parasite Light micrograph of a cercaria of the trematode* Schistosoma haematobium, *the cause of the disease schistosomiasis (bilharzia). This cercarial stage lives in lakes and rivers, and penetrates the skin of humans in the water. It matures to the adult form, then migrates to the veins surrounding the bladder. Here the females lay eggs which pass out in the urine and contaminate the local water supply.*

**P**

facial expression. It arises from a lack of the NEUROTRANSMITTER, dopamine. Slightly more common in men, it is rare before the age of 50. Foremost among the drugs used to control the disease is L-DOPA.

**pathology** Study of diseases, their causes and the changes they produce in the cells, tissues, and organs of the body.

**pediatrics** Medical specialty devoted to the diagnosis and treatment of disease and injury in children. Specialists in this field require a thorough knowledge not only of a wide range of conditions peculiar to children, but also of normal childhood development and the ways in which it may affect treatment and recovery.

**pellagra** Disease caused by a deficiency of nicotinic acid, one of the B group of VITAMINS. Its symptoms are lesions of the skin and mucous membranes, diarrhea, and mental disturbance.

**penicillin** ANTIBIOTIC agent derived from molds of the genus *Penicillium*. Sir Alexander Fleming discovered penicillin, the first antibiotic, in 1928. It was synthesized and first became available in 1941. Penicillin was widely used for treating casualties in World War II. It can produce allergic reactions, and some microorganisms have become resistant.

**pepsin** Digestive enzyme secreted by glands of the stomach wall as part of the GASTRIC JUICE. In the presence of hydrochloric acid, it catalyzes the splitting of proteins in food into polypeptides. *See* pages 86–87

**peptic ulcer** ULCER involving those areas of the digestive tract exposed to PEPSIN. Peptic ulcers most commonly occure in the stomach (**gastric ulcer**) and the first part of the duodenum (**duodenal ulcer**). Duodenal ulcers are more common than gastric ulcers and usually occur in people aged 20 to 45.

**peristalsis** Series of wavelike movements that propels food through the gut or digestive tract. It is caused by contractions of the smooth involuntary muscle of the gut wall. The reverse process, antiperistalsis, produces vomiting.

**peritonitis** Inflammation of the peritoneum – the strong connective tissue lining the abdominal walls and organs. It may be caused by bacterial infection or chemical irritation or it may arise spontaneously in certain diseases. Symptoms include fever, abdominal pain and distension, and shock. Treatment is directed at the underlying cause.

**phagocyte** Type of LEUKOCYTE able to engulf other cells, such as bacteria. Part of the body's IMMUNE SYSTEM, it digests what it engulfs in the defense of the body against infection. Phagocytes also act as scavengers by clearing the bloodstream of the remains of the cells that die as part of the body's natural processes.

P

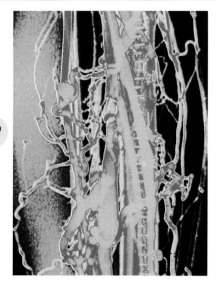

**pharmacology** Study of the properties of DRUGS and their effects on the body.

**pharmacopeia** Reference book listing drugs and other preparations in medical use. Included are details of their formulas, dosages, routes of administration, known side effects, and precautions.

**phlebitis** Inflammation of the wall of a vein. It may be caused by infection, trauma, underlying disease, or the presence of VARICOSE VEINS. Symptoms include localized swelling and redness. Treatment includes rest and anticoagulant therapy.

**physical therapy** Use of various physical techniques to treat disease or injury. Its techniques include massage, manipulation, exercise, heat, hydrotherapy, the use of ultrasonics, and electrical stimulation.

**physiology** Branch of biology concerned with the functions of living organisms, as opposed to their structure (anatomy).

**pill, the** Popular term for oral contraceptives based on the female reproductive HORMONES. They work by preventing ovulation. Two types of synthetic hormone,

◀ *Phlebitis Colored X-ray showing the inflammation of veins (known as phlebitis) in the leg of a patient. The veins have been colored orange and appear misshapen, leading off randomly (such as at left of image).*

similar to ESTROGEN and PROGESTERONE, are generally used, though the former alone is effective in preventing ovulation; the latter helps to regulate the menstrual cycle. Possible side effects include headache, HYPERTENSION, weight gain, and a slightly increased risk of THROMBOSIS. *See also* CONTRACEPTION

**plague** Acute, infectious disease of humans and rodents caused by the bacillus *Yersinia pestis*. In humans, it occurs in three forms: **bubonic** plague, the most common and characterized by vomiting, fever and swellings of the lymph nodes called 'buboes'; **pneumonic** plague, in which the lungs are infected; and **septicemic** plague, in which the bloodstream is invaded. Treatment is the administration of vaccines, bed rest, antibiotics, and sulfa drugs. *See also* BLACK DEATH

**plasma** In biology, liquid portion of the BLOOD. It contains an immense number of ions, inorganic and organic molecules such as immunoglobulins, and hormones and their carriers but not cells. It clots upon standing.

**plastic surgery** Branch of surgery that involves the reconstruction of deformed, damaged, or disfigured parts of the body. **Cosmetic surgery**, such as face lifts, is performed solely to 'improve' appearance.

**platelet** Colorless, usually spherical structures found in mammalian BLOOD. Chemical compounds in platelets, known as factors and cofactors, are essential to the mechanism of blood clotting. The normal platelet count is c.300,000 per cu mm of blood.

**pleurisy** Inflammation of the pleura – the membrane that lines the space between the lungs and the chest walls. It is nearly always due to infection, but may arise as a complication of other diseases.

**pneumoconiosis** Occupational disease of workers, principally miners, in confined and dusty conditions. Caused by inhaling irritants, often only as minute specks, the disease inflames and can finally destroy lung tissue.

**pneumonia** Inflammation of the lung tissue, most often caused by bacterial infection. Most at risk are the very young, the aged, and those whose immune systems have been undermined by disease or certain medical treatments. The most common form is pneumococcal pneumonia, caused by the bacterium *Streptococcus pneumoniae*. Symptoms include fever, chest pain, coughing, and the production of rust-colored sputum. Treatment is with ANTIBIOTICS.

**podiatry** Treatment and care of the foot. Podiatrists treat such conditions as corns and bunions and devise ways to accommodate foot deformities.

**poliomyelitis** (polio) Acute viral infection of the nervous system affecting the nerves that activate muscles. Often a mild disease with effects limited to the throat

and intestine, it is nonetheless potentially serious, with paralysis occurring in 1% of patients. It becomes life-threatening only if the breathing muscles are affected, in which case the person may need artificial ventilation. It has become rare in developed countries since the introduction of vaccination in the mid-1950s.

**porphyria** Group of rare genetic disorders in which there is defective METABOLISM of one or more porphyrins, the breakdown products of HEMOGLOBIN. It can produce a wide range of effects, including intestinal upset, HYPERTENSION, weakness, abnormal skin reactions to sunlight, and mental disturbance. A key diagnostic indicator is that the patient's urine turns reddish-brown if it is left to stand.

**positron emission tomography (PET)** Medical imaging technique (used particularly on the brain) that produces three-dimensional images. Radioisotopes, injected into the bloodstream prior to imaging, are taken up by tissues where they emit positrons that produce detectable photons. *See* picture, page 168

**post-mortem** (autopsy) Dissection of a body to determine the cause of death. It is performed to confirm a diagnosis or to

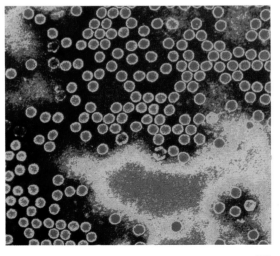

▶ *Poliomyelitis TEM of clusters of polio viruses, the cause of poliomyelitis. There are three serotypes of the polio virus. The serotype shown here is type 1, cause of most epidemics. Although rounded in shape (red) each virus has a cubic symmetry. They are picorna-viruses. RNA genetic material occurs in the core of each virus, surrounded by a protein coat (yellow). Infection occurs through the mouth. Usually only a mild illness results, but in more serious cases the virus may attack the nervous system causing paralysis or even death. Polio vaccines are available.*

**P**

establish the cause of an unexpected death. **Morbid anatomy** (the examination of the dead) is a branch of PATHOLOGY.

**post-natal depression** Mood disorder, characterized by intense sadness, which may develop in a mother within a few days of childbirth. It ranges from mild cases of the 'baby blues,' which are usually short-lived, to the severe depressive illness known as **puerperal psychosis**.

**post-traumatic stress disorder** Anxiety condition that may develop in people who have been involved in or witnessed some horrific event. It is commonly seen in survivors of battles or major disasters. The condition is characterized by repeated flashbacks to distressing events, hallucinations, nightmares, insomnia, edginess, and depression. It usually recedes over time, but as many as 10% of sufferers are left with permanent psychological disability.

**pregnancy** Period of time from conception until birth, in humans normally c.40 weeks (280 days). It is generally divided into three 3-month periods called trimesters. In the **first trimester**, the embryo grows from a small ball of cells to a fetus c.3 in (7.6 cm) in length. At the beginning of the **second trimester**, movements are first felt and the fetus grows to about 14 in (36 cm). In the **third trimester**, the fetus attains its full body weight. See also LABOR; Pregnancy and Childbirth, pages 104–105

**premature birth** Birth of a baby prior to 37 weeks' gestation or weighing less than 5.5 lb (2.5 kg). Premature babies are more at risk than those born at full term and require special care.

**prescription** Order written by a doctor for drugs or other medication to be dispensed by a pharmacist. It should give details of the quantity of drugs to be dispensed, the dosage required, the route of administration (such as by mouth), and any precautions.

**prickly heat** Skin rash caused by blockage of the sweat glands in hot, humid weather. It occurs most often in infants and obese people. It disappears as the body cools.

**progesterone** Steroid hormone secreted mainly by the corpus luteum of the mammalian ovary and by the placenta during pregnancy. Its principal function is to prepare and maintain the inner lining (ENDOMETRIUM) of the uterus for pregnancy. Synthetic progesterone is one of the main components of the contraceptive PILL.

**prolactin** Hormone, secreted by the anterior pituitary gland, which stimulates the production of milk after childbirth. It also stimulates the secretion of proges-

◀ *Positron Emission Tomography (PET) PET scans detect blood flow. These PET scans show areas of the brain activated by different tasks. At upper left, sight activates the visual area in the occipital cortex. Upper right, hearing activates the auditory area in the superior temporal cortex. Lower left, speaking activates the insula and motor cortex. Lower right, thinking about verbs and speaking them generates high activity in hearing, speaking, temporal, and parietal areas.*

► *Progesterone* PLM of crystals of progesterone. Progesterone is the most potent of progestogens, a class of steroid hormones. It derives from pregnenolone. The primary function of progesterone is to prepare the inner lining of the womb for pregnancy. It is secreted by the ovaries after the release of an egg. If this egg is fertilized, the ovaries produce progesterone for four months, after which it is secreted by the placenta.

**P**

terone by the corpus luteum of the ovary. *See also* pages 56–57 and pages 106–107

**prolapse** Displacement of an organ due to weakening of supporting tissues. It most often affects the rectum, due to bowel problems, or the uterus following repeated pregnancies.

**prostaglandin** Series of related fatty acids, with hormonelike action, present in SEMEN and liver, brain, and other tissues. Their biological effects include the lowering of blood pressure and the stimulation of contraction in a variety of smooth-muscle tissues, such as in the uterus.

**prosthesis** Artificial substitute for a missing organ or part of the body. Until the 17th century, artificial limbs were made of wood or metal, but innovations in metallurgy, plastics, and engineering enabled lighter, jointed limbs to be made. More recent prosthetic devices include artificial heart valves made of silicone materials.

**protein** Organic compound containing many amino acids linked together by covalent, peptide bonds. Living cells use *c.*20 different amino acids, which are present in varying amounts. The GENETIC CODE, carried by the DNA of the CHROMOSOMES, determines which amino acids are used and in which order they are combined. The most important proteins are enzymes, which determine all the chemical reactions in the cell, and antibodies, which combat infection. **Structural** proteins include KERATIN and COLLAGEN. **Gas** transport proteins include HEMOGLOBIN. Nutrient proteins include casein. Protein HORMONES regulate the METABOLISM. *See also* ANTIBODY

**Prozac** Trade name for one of a small group of antidepressants, known as **selective serotonin re uptake inhibitors** (SSRIs). They work by increasing levels of **serotonin** in the brain. Serotonin, or 5-hydroxytryptamine (5-HT), is a NEUROTRANSMITTER involved in a range of functions. Low levels of serotonin are associated with depression.

**psittacosis** (parrot fever) Disorder usually affecting the respiratory system of birds. Caused by a bacterium, it can be transmitted to human beings, producing pneumonialike symptoms. Treatment is with ANTIBIOTICS.

**psoriasis** Chronic recurring skin disease featuring raised, red, scaly patches. The lesions frequently appear on the chest, knees, elbows, and scalp. Treatment is with tar preparations, steroids, and ultraviolet light. Psoriasis is sometimes linked with a form of ARTHRITIS.

**psychosis** Serious mental disorder in which the patient loses contact with reality, in contrast to NEUROSIS. It may feature extreme mood swings, delusions or hallucinations, distorted judgment, and inappropriate emotional responses. **Organic** psychoses

P

Q

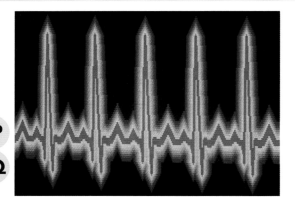

◀ *Pulse* Computer graphics abstract based on a section of the trace of a human pulse, the beat of the heart as it pumps blood around the body. Five successive beats are shown in this image.

may spring from brain damage, advanced SYPHILIS, senile DEMENTIA, or advanced EPILEPSY. **Functional** psychoses, for which there is no known organic cause, include schizophrenia and MANIC DEPRESSION.

**psychosomatic disorder** Physical complaint thought to be rooted, at least in part, in psychological factors. The term has been applied to many complaints, including asthma, migraine, ulcers, and hypertension.

**puberty** Time in human development when sexual maturity is reached. The reproductive organs take on their adult form, and secondary sexual characteristics, such as the growth of pubic hair, start to become evident. Girls develop breasts and begin to menstruate; in boys there is deepening of the voice and the growth of facial hair. Puberty may begin at any time from about the age of ten, usually occurring earlier in girls than in boys. Hormones known as gonadotrophins regulated the process. *See also* pages 108–109

**pulse** Regular wave of raised pressure in arteries that results from the flow of BLOOD pumped into them at each beat of the heart. The pulse is usually taken at the wrist, although it may be observed at any point where an artery runs close to the body surface. The average pulse rate is c.70 per minute in adults.

**purpura** Purplish patches on the skin due to seepage of blood from underlying vessels. Physical injury, vitamin deficiencies, some allergies, and certain drugs are among the causes.

**pus** Yellowish fluid forming as a result of bacterial infection. It comprises blood serum, LEUKOCYTES, dead tissue, and living and dead BACTERIA. An ABSCESS is a pus-filled cavity.

**pyelitis** Inflammation of the pelvis of the kidney, where urine collects before draining into the ureter. More common in women, it is usually caused by bacterial infection. Treatment is with ANTIBIOTICS and copious fluids.

**quinine** White, crystalline substance isolated in 1820 from the bark of the cinchona tree. It was once widely used in the treatment of MALARIA but has been largely replaced by drugs that are less toxic and more effective.

**quinsy** Inflammation of the tonsils, caused by an abscess, often a complication of TONSILLITIS. It is generally treated with ANTIBIOTICS.

**rabies** (hydrophobia) Viral disease of the central nervous system. It can occur in all warm blooded animals, but is especially feared in dogs due to the risk of transmission to human beings. The incubation period varies from a week or two to more than a year. It is characterized by severe thirst, although attempting to drink causes painful spasms of the larynx; other symptoms include fever, muscle spasms, and delirium. Once the symptoms appear, death usually follows within a few days. Anyone bitten by a rabid animal may be saved by prompt injections of rabies vaccine and antiserum.

**radiation sickness** Illness resulting from exposure to sources of ionizing radiation, such as X-rays, gamma rays, or nuclear fallout. Diarrhea, vomiting, fever, and hemorrhaging are symptoms. Severity depends upon the degree of radiation, and treatment is effective in mild cases.

**radiology** Medical specialty concerned with the use of radiation and radioactive materials in the diagnosis and treatment of disease. *See also* RADIOTHERAPY

▶ *Reflexology A chart indicating reflexology points on the foot. Areas of the foot that affect different parts of the body are colored and numbered. Reflexology is a healing therapy originating from "zone theory".*

**radiotherapy** Use of radiation to treat TUMORS or other pathological conditions. It may be done either by implanting a pellet of a radioactive source in the part to be treated, or by dosing the patient with a radioactive isotope or by exposing the patient to precisely focused beams of radiation from a machine such as an X-ray machine or a particle accelerator. Cobalt-60 is often used as it produces highly penetrating gamma radiation. In the treatment of CANCERS, the radiation slows down the proliferation of the cancerous cells.

**reflex action** Rapid involuntary response to a particular stimulus – for example, the 'knee-jerk' reflex that occurs when the bent knee is tapped. It is controlled by the nervous system.

**R**

**reflexology** School of complementary medicine based on the theory that the image of the body is reflected in the foot. Modern practices are based on theories that date back as far as 3000 BC. Practitioners use foot massage to clear blockages of energy flow in the body, which cause illness.

**repetitive strain injury (RSI)** Pain and reduced mobility in a limb, most often the wrist, caused by constant repetition of the

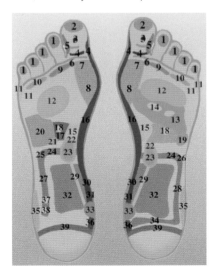

R

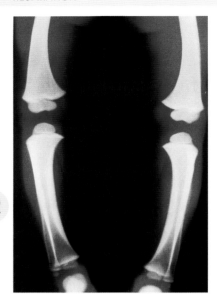

◀ *Rickets* Colored X-ray of the weakened bones and bowed legs of a child suffering from rickets. The skeleton of the legs appears deformed, with the long bones of the limbs severely curved. Rickets is caused by a nutritional deficiency of vitamin D. In adults, this condition is known as osteomalacia.

**resuscitation** Measures taken to revive a person who is on the brink of death. The most successful technique available to the layman is **mouth-to-mouth** resuscitation. Medical staff receive instruction in **cardiopulmonary** resuscitation (CPR), which involves the use of specialized equipment and drugs to save patients whose breathing and/or heartbeat suddenly stop. *See also* FIRST AID

**retrovirus** Any of a large VIRUS family (Retroviridae) that, unlike other living organisms, contains the genetic material RNA (ribonucleic acid) rather than the customary DNA (deoxyribonucleic acid). In order to multiply, retroviruses make use of a special enzyme to convert their RNA into DNA, which then becomes integrated with the DNA in the cells of their hosts. Diseases caused by retroviruses include ACQUIRED IMMUNE DEFICIENCY SYNDROME (AIDS).

**rheumatic fever** Inflammatory disorder characterized by fever and painful swelling of the joints. Rare in the modern developed world, it mostly affects children and young adults. An important complication is possible damage to the heart valves, leading to rheumatic heart disease in later life.

**rheumatism** General term for a group of disorders whose symptoms are pain, inflammation, and stiffness in the bones, joints, and surrounding tissues. Usually some form of ARTHRITIS is involved.

**rhinitis** Inflammation of the mucous membrane of the nose. It may be an allergic reaction (such as HAY FEVER), or a symptom of a viral infection such as the common cold.

**rickets** Disorder in which there is defective growth of bone in children; the bones fail to harden sufficiently and become

same movements. The symptoms arise from inflammation of the tendon sheaths because of excessive use. RSI is an occupational disorder mostly seen in assembly-line workers and keyboard operators.

**respiration** Series of chemical reactions by which complex molecules (food molecules) are broken down to release energy in living organisms. Enzymes control these reactions, which are an essential part of METABOLISM. There are two main types of respiration: aerobic and anaerobic. In **aerobic** respiration, oxygen combines with the breakdown products and is necessary for the reactions to take place. **Anaerobic** respiration takes place in the absence of oxygen. In most living organisms, the energy released by respiration is used to convert **adenosine diphosphate** (ADP) to **adenosine triphosphate** (ATP), which transports energy around the cell. At the site where the energy is needed, ATP converts back to ADP with the aid of a special enzyme and energy releases. The first stages of respiration take place in the cytoplasm and the later stages in the mitochondria. *See also* Respiratory system, pages 74–79

bent. Due either to a lack of vitamin D in the diet or to insufficient sunlight to allow its synthesis in the skin, it results from the inability of the bones to calcify properly.

**rigor mortis** Stiffening of the body after death brought about by chemical changes in muscle tissue. Onset is gradual from minutes to hours. Rigor mortis disappears within about 24 hours after death.

**ringworm** Fungus infection of the skin, scalp, or nails. The commonest type of ringworm is athlete's foot (*tinea pedis*). It is treated with antifungal preparations.

**RNA** (ribonucleic acid) Chemical (nucleic acid) that controls the synthesis of PROTEIN in a cell and is the genetic material in some viruses. The molecules of RNA in a cell are copied from DNA and consist of a single strand of nucleotides, each containing the sugar ribose, phosphoric acid, and one of four bases: adenine, guanine, cytosine, or uracil. **Messenger RNA (mRNA)** carries the information for protein synthesis from DNA in the cell nucleus to the ribosomes in the cytoplasm. **Transfer RNA** brings amino acids to their correct position on the messenger RNA. Each amino acid is specified by a sequence of three bases in mRNA.

**roundworm** Parasite of the class Nematoda, which inhabits the intestine of mammals. It breeds in the intestine. The larva bores through the intestinal wall, is carried to the lungs in the bloodstream, and crawls to the mouth where it is swallowed. Length: 6–12 in (15–30 cm).

**RSI** Abbreviation of REPETITIVE STRAIN INJURY

**rubella** *See* GERMAN MEASLES

**rupture** *See* HERNIA

▶ *Salmonella Colored TEM of a group of* Salmonella *sp. bacteria. Each cell is rod-shaped (blue/green) and has long hairlike flagella (yellow) used for motility. Salmonella bacteria are of the Enterobacteriaceae family. They are gram-negative bacilli bacteria that inhabit the gut. More than 1,800 different serotypes of Salmonella are known.*

**saliva** Fluid secreted into the mouth by the salivary glands. In vertebrates, saliva consists of c.99% water with dissolved traces of sodium, potassium, calcium, and the enzyme amylase. Saliva softens and lubricates food to aid swallowing, and amylase starts the digestion of starches. *See also* pages 84–85

**salmonella** Several species of rod-shaped BACTERIA that cause intestinal infections in human beings and animals. *Salmonella typhi* causes TYPHOID FEVER; other species cause GASTROENTERITIS. The bacteria are transmitted by carriers, particularly flies, and in food and water.

**sarcoma** Cancerous growth or TUMOR arising from muscle, fat, bone, blood or

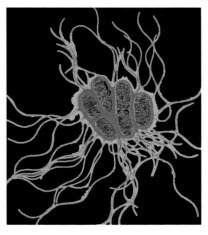

lymph vessels, and connective tissue. *See also* CANCER

**scabies** Contagious infection caused by a female mite, *Sarcoptes scabiei*, which burrows into the skin to lay eggs. It can be seen as a dark wavy line on the skin and is treated with antiparasitic creams.

**scanning** In medicine, use of a noninvasive system to detect abnormalities of structure or function in the body. Detectable waves (X-rays, gamma rays, ultrasound) pass through the part of the body to be investigated and the computer-analyzed results displayed as images on a viewing screen.

**scarlet fever** (scarlatina) Acute infectious disease, usually affecting children, caused by BACTERIA in the *Streptococcus pyogenes* group. It is characterized by a bright red body rash, fever, vomiting, and a sore throat. It is treated with ANTIBIOTICS.

**schistosomiasis** (bilharzia) Visceral, venous infestation of the human body by blood flukes of the genus *Schistosoma* occurring mainly in the tropics. Next to MALARIA, it is the most serious parasitic human infection. Symptoms are skin eruption, inflammation, fever, and often swelling of the liver. It is contracted from water contaminated with microscopic larvae released by snail hosts. The larvae enter the body through the skin, mature in the blood, and deposit eggs throughout the body. Treatment is with drugs containing antimony or with a course of chemotherapy.

**schizophrenia** Severe mental disorder marked by disturbances of cognitive functioning, particularly thinking. As well as the characteristic loss of contact with reality, symptoms can include hallucinations and delusions, and muffled or inappropriate emotions. Research suggests that schizophrenia may be caused by high levels of dopamine, a NEUROTRANSMITTER.

**sciatica** Severe pain in the back and radiating down one or other leg, along the course of the sciatic nerve. It is usually caused by inflammation of the sciatic nerve or by pressure on the spinal nerve roots.

**sclerosis** Degenerative hardening of tissue, usually due to scarring following inflammation or as a result of aging. It can affect the brain and spinal cord, causing neurological symptoms, or the walls of the arteries.

**screening** In medicine, a test applied either to an individual or to groups of people who, though apparently healthy, are known to be at special risk of developing a particular

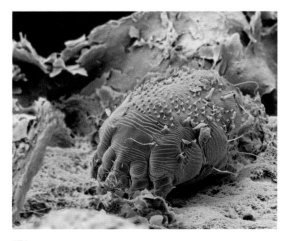

◄ *Scabies Colored SEM of an itch mite,* Sarcoptes sp., *on the surface of a piece of skin. This parasite attacks human skin, causing scabies. Scabies is typified by severe itching (especially at night), red papules, and often secondary infection. The female mite tunnels in the skin to lay her eggs and the newly hatched mites are passed easily from person to person by physical contact. The itching is caused by an allergic reaction to the mite's saliva or feces. Magnification: ×110.*

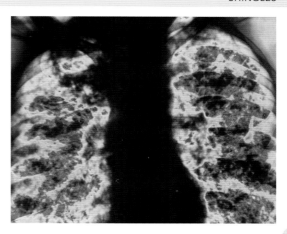

▶ *Silicosis False-color X-ray of the chest area of a person suffering from silicosis, a lung disease caused by inhaling silica dust particles. The silica stimulates fibrosis of lung tissue (apparent as blue areas over both left and right lungs), which results in breathlessness and an increased susceptibility to tuberculosis. Workers in the mining, quarrying, stone dressing, sandblasting and boiler scaling industries are prone to the disease.*

disease. It is used to detect treatable diseases while they are still in the early stages.

**scurvy** Disease caused by a deficiency of vitamin C (ascorbic acid), which is contained in fresh fruit and vegetables. It is characterized by weakness, painful joints, and bleeding gums.

**seasonal affective disorder (SAD)** Mental depression apparently linked to the seasonally changing amount of light. Sufferers experience depression during fall and winter.

**seborrhea** Disorder of the sebaceous glands characterized by overproduction of sebum that results in red, scaly patches on the skin and dandruff.

**sedative** Drug used for its calming effect, to reduce anxiety and tension; at high doses it induces sleep. Sedatives include NARCOTICS, BARBITURATES, and BENZODIAZEPINES.

**semen** Fluid in a male that contains sperm from the testes and the secretions of various accessory sexual glands. In men, each ejaculate is normally *c.*1 to 2 oz (3 to 6 ml) by volume, and contains about 200 to 300 million sperm. *See also* pages 98–99

**sepsis** Destruction of body tissue by disease-causing (pathogenic) bacteria or their toxins. Local or widespread inflam-

mation may occur, possibly followed by NECROSIS, the death of tissue. Treatment is with ANTIBIOTICS.

**septicemia** Term for severe SEPSIS or BLOOD POISONING

**serum** Clear fluid that separates out if blood is left to clot. It is essentially of the same composition as PLASMA, but without fibrinogen and clotting factors.

**sexual intercourse** Term used to describe sexual relations between people. The term is most commonly used to describe the insertion of a man's penis into a woman's vagina. *See also* CONTRACEPTION; ORGASM; Reproductive system, pages 98–103

**sexually transmitted disease (STD)** Any disease that is transmitted by sexual activity involving the transfer of body fluids. It encompasses a range of conditions that are spread primarily by sexual contact, although they may also be transmitted in other ways. These include ACQUIRED IMMUNE DEFICIENCY SYNDROME (AIDS), pelvic inflammatory disease, cervical cancer, and viral HEPATITIS. Older STDs, such as SYPHILIS and GONORRHEA, remain significant public health problems.

**shingles** (herpes zoster) Acute viral infection of sensory nerves. Groups of small blisters appear along the course of the

S

S

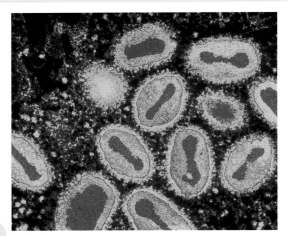

◀ Smallpox Colored TEM of a section through many variola major viruses. These agents cause smallpox. The protein coat of each virus is colored yellow; DNA genetic material is red. The viruses display a complex symmetry. The disease causes high fever and skin spots that quickly develop into scarring pustules. The virus is transmitted by coughed droplets or the pus from pustules of an infected person. The virus is now used for purposes of biological research. Magnification: ×28,500.

affected nerves, and the condition can be very painful.

**shock** Acute circulatory failure, possibly with collapse. Caused by disease, injury, or emotional trauma, it is characterized by weakness, pallor, sweating, and a shallow, rapid pulse. In shock, blood pressure drops below the level needed to oxygenate the tissues.

**Siamese twins** Identical twins who are born physically joined together, sometimes with sharing of organs. Surgical separation is usually possible.

**sickle-cell anemia** Inherited blood disorder featuring an abnormalcy of HEMOGLOBIN. The hemoglobin is sensitive to a deficiency of oxygen and it distorts ERYTHRO-CYTES (red blood cells), causing them to become rigid and sickle-shaped. Sickle cells are rapidly lost from the circulation, giving rise to anemia and jaundice.

**sight** Sense by which form, color, size, movement, and distance of objects are perceived. Essentially, it is the detection of light by the eye, enabling the formation of visual images. See also pages 50–53

**silicosis** Chronic lung disease, caused by prolonged inhalation of silica dust in occupations such as mining and stone grinding. See picture, page 175

**sleep** Periodic state of unconsciousness from which a person or animal can be roused. During an ordinary night's sleep, there are intervals of deep sleep associated with rapid eye movement (REM) sleep. It is during this REM sleep that dreaming occurs. Studies have shown that people deprived of sleep become grossly disturbed. Sleep requirement falls sharply in old age. Difficulty in sleeping is called INSOMNIA. See also DREAM

**sleeping sickness** Popular name of two diseases. **Trypanosomiasis** is a disease of tropical Africa caused by a parasite transmitted by the tsetse fly. It is characterized by fever, headache, joint pains, and anemia. Ultimately, it may affect the brain and spinal cord, leading to profound lethargy and sometimes death. **Encephalitis lethargica** is a rare, viral disease of the brain characterized by headache and drowsiness progressing to coma. It occurs in epidemic and sporadic forms and, most notoriously, was the cause of an epidemic that followed World War I, leaving many helpless survivors.

**slipped disk** (prolapsed intervertebral disk) Protrusion of the soft, inner core of an intervertebral disk through its covering, causing pressure on the spinal nerve roots. It is caused by a sudden mechanical force on the spine. It causes stiffness and

SCIATICA. Initial treatment is with strong painkillers. Physiotherapy is often used.

**smallpox** Formerly a highly contagious viral disease characterized by fever, vomiting, and skin eruptions. It remained endemic in many countries until the World Health Organization (WHO) campaign, launched in the late 1960s. Global eradication was achieved by the early 1980s. As late as the 18th century, smallpox killed one in 10 children born in France.

**smell** (olfaction) Sense that responds to airborne molecules. The **olfactory receptors** in the nose can detect even a few molecules per million parts of air. *See also* Taste and smell, pages 48–49

**snakebite** Injection of potentially lethal snake venom into the bloodstream. There are three types of venomous snake: the **Viperidae**, subdivided into true vipers and pit vipers, whose venom causes internal hemorrhage; the **Elapidae** (including cobras, mambas, kraits), whose venom paralyzes the nervous system; and the **Hydrophidae**, Pacific sea snakes with venom that disables the muscles. Treatment for all three types is with antivenoms.

**sodium bicarbonate** (sodium hydrogen carbonate, $NaHCO_3$, popularly known as bicarbonate of soda) White, crystalline salt that decomposes in acid or on heating to release carbon dioxide gas. It has a slightly alkaline reaction and is an ingredient of indigestion medicines.

**spasm** Sustained involuntary muscle contraction. It may occur in response to pain, or as part of a generalized condition, such as spastic paralysis or TETANUS.

**sphygmomanometer** Instrument used to measure BLOOD PRESSURE. The instrument incorporates an inflatable rubber cuff connected to a column of mercury with a graduated scale. The cuff wraps around the upper arm and inflates to apply tension to a major artery. When the air slowly releases, blood pressure readings can be ascertained from the scale.

**spina bifida** Congenital disorder in which the bones of the spine do not develop properly to enclose the spinal cord. Surgery to close the defect is usually performed soon after birth, but this may not cure disabilities caused by spina bifida. Taking folic acid in early pregnancy reduces the risk.

**S**

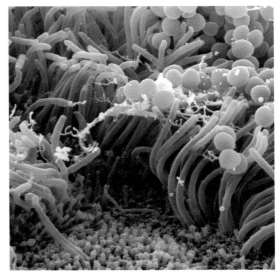

▶ *Staphylococcus Colored SEM of* Staphylococcus aureus *bacteria (yellow) on human nasal epithelial cells. These gram-positive cocci (spherical bacteria) are adhering to mucus (blue) on the hairlike cilia which protrude from the epithelial cells.* S. aureus *is very common in humans, living harmlessly on the skin and in the nose, throat, and large intestine. However it can multiply inside blocked skin pores, causing pus to build up and forming an acne spot or boil. Inside the body this bacteria may cause abscesses and suppurative infection. Magnification: ×7,000.*

**spinal tap** (lumbar puncture) Procedure for withdrawing CEREBROSPINAL FLUID from the lumbar (lower back) portion of the spinal cord for laboratory examination to aid diagnosis.

**sprain** Injury to one or more ligaments of a joint caused by sudden overstretching. Symptoms include pain, stiffness, bruising, and swelling. Treatment includes resting and supporting the affected part before gentle mobilization.

**squint** See STRABISMUS

**staphylococcus** Spherical bacterium that grows in grapelike clusters and is found on the skin and mucous membranes of human beings and other animals. Pathogenic staphylococci cause a range of local or generalized infections, including PNEUMONIA and SEPTICEMIA. They may be destroyed by ANTIBIOTICS, although some strains have become resistant. See artwork, page 177

**sterility** Inability to reproduce. It may be due to INFERTILITY or, in humans and other animals, to STERILIZATION.

**sterilization** Surgical intervention that terminates the ability of a human or other animal to reproduce. In women, the usual procedure is tubal ligation: sealing or tying off the Fallopian tubes so that fertilization can no longer take place. In men, a VASECTOMY is performed to block the release of sperm. The term is also applied to the practice of destroying microorganisms in order to prevent the spread of infection. Techniques include heat treatment, irradiation, and the use of disinfecting agents. See also pages 98–103

**steroid** Class of organic compounds with a basic molecular structure of 17 carbon atoms arranged in four rings. Steroids are widely distributed in animals and plants, the most abundant being the sterols, such as cholesterol. Another important group are the steroid hormones, including the corticosteroids, secreted by the adrenal cortex, and the sex hormones (ESTROGEN, PROGESTERONE, and TESTOS-TERONE). Synthetic steroids are widely used in medicine. Athletes sometimes abuse steroids to increase their muscle mass, strength, and stamina, but there are harmful side effects. Taking steroids is illegal in sports.

**stethoscope** Instrument that enables an examiner to listen to the action of various parts of the body, principally the heart and lungs. It consists of two earpieces attached to flexible rubber tubes that lead to either a disk or a cone.

**stimulant** Substance that increases mental alertness and activity. There are a number of stimulants that act on the central nervous system, notably drugs in the AMPHETAMINE group. Many common beverages, including tea and coffee, contain small quantities of the stimulant caffeine.

**strabismus** (squint) Condition in which the eyes do not look in the same direction. It may result from either disease of or damage to the eye muscles or their nerve supply, or an error of refraction within the eye.

**streptococcus** Genus of gram-positive spherical or oval BACTERIA that grow in pairs or beadlike chains. They live mainly as PARASITES in the mouth, respiratory tract and intestine. Some are harmless but others are pathogenic, causing SCARLET FEVER and other infections. Treatment is with ANTIBIOTICS.

**stroke** (apoplexy) Interruption of the flow of blood to the brain. It is caused by blockage or rupture of an artery and may produce a range of effects from mild impairment to death. Conditions that predispose to stroke include ARTERIOSCLEROSIS and HYPERTENSION. Many major strokes are prevented by treatment of risk factors, including surgery and the use of anticoagulant drugs. **Transient ischemic attacks** (TIAs), or 'mini-strokes,' which last less than 24 hours, are investigated to try to prevent the occurrence of a more damaging stroke.

**strychnine** Poisonous alkaloid obtained from the plant Strychnos nux-vomica. In the past, it was believed to have therapeutic value in small doses as a tonic. Strychnine poisoning causes symptoms similar to those of TETANUS, with death occurring due to SPASM of the breathing muscles.

**sudden infant death syndrome (SIDS)** (Crib death) Unexpected death of an apparently healthy baby, usually during sleep. The peak period seems to be around two months' old, although it can occur up to a year or more after birth. Claiming more boys than girls, it is more common in winter. The cause is unknown, but a number of risk factors have been identified, including prematurity and respiratory infection.

**sulfonamide drug** Any of a group of DRUGS derived from sulfanilamide, a red textile dye, that prevents the growth of bacteria. Introduced in the 1930s, sulfon-amides were the first antibacterial drugs, pre-scribed to treat a range of infections. They were replaced by less toxic and more effec-tive ANTIBIOTICS.

**sunburn** Damage to skin caused by pro-longed or unaccustomed exposure to sunlight. It varies in severity from redness and soreness to the formation of large blisters. Excessive exposure to sunlight is associated with the skin CANCER known as **melanoma**.

**sunstroke** Potentially fatal condition caused by overexposure to direct sunlight, in which the body temperature rises to 105°F (40.5°C) or more. Symptoms include hot, dry skin, exhaustion, vomiting, delirium, and coma. Urgent medical treatment is required, possibly in an intensive care unit. Recovery is usual within a day or two.

**surgery** Branch of medical practice concerned with treatment by operation. Traditionally, it has mainly involved open surgery: gaining access to the operative site by way of an incision. However, the practice of using ENDOSCOPES enabled the development of 'keyhole surgery', using minimally invasive techniques. Surgeons perform operations under sterile condi-tions using local or general ANESTHESIA.

**syncope** *See* FAINTING

**synovial fluid** Viscous, colorless fluid that lubricates the movable joints between bones. It is secreted by the synovial membrane. Synovial fluid is also found in the **bursae**, membranous sacs that help to reduce friction in major joints.

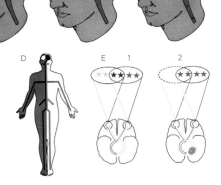

▶ *Stroke A 'stroke' is a disruption of the blood supply to the brain and is due either to a hemorrhage or a cerebral blood pressure (A), a thrombus (B), or an embolus (C). The damage (stippled area) may be permanent, but the outcome of a stroke depends on the extent and area of the brain affected by it. The symptons can range from a temporary loss of speech or other brain function and paralysis of the limbs to sudden death. A stroke on one side of the brain affects the limbs of the opposite side of the body because of the crossing over of nerve tracts in the brain stem (D). There is a similar effect on the visual cortex (E). Normal vision (1) is impaired on the left side (2) by a stroke on the right side.*

**syphilis** SEXUALLY TRANSMITTED DISEASE caused by the spiral-shaped bacterium (spirochete) *Treponema pallidum*. Untreated, it runs its course in three stages. The first symptom is often a hard, painless sore on the genitals, appearing usually within a month of infection. Months later, the second stage features a skin rash and fever. The third stage, often many years later, brings the formation of growths and serious involvement of the heart, brain, and spinal cord, leading eventually to blindness, insanity, and death. The disease is treated with ANTIBIOTICS.

**T**

**tachycardia** Increase in heart rate beyond the normal. It may occur after exertion or because of excitement or illness, particularly during fever; or it may result from a heart condition.

**tapeworm** Parasite of the genus *Taenia*, which colonizes the intestines of vertebrates, including human beings. Caught from eating raw or undercooked meat, it may cause serious disease.

**tears** Salty fluid secreted by glands that moistens the surface of the eye. It cleanses and disinfects the surface of the eye and also brings nutrients to the cornea.

**temperature** In biology, intensity of heat. In warm blooded (homeothermic) animals, body temperature is maintained within narrow limits regardless of the temperature of their surroundings. This is accomplished by muscular activity, the operation of cooling mechanisms, such as vasodilation, vasoconstriction and sweating, and metabolic activity. In humans, normal body temperature is c.98.4°F (36.9°C), but this varies with the degree of activity. In cold blooded (poikilothermic) animals, body temperature varies more widely, depending on the temperature of the surroundings.

**testosterone** STEROID hormone secreted mainly by the mammalian testis. It is responsible for the growth and development of male sex organs and male secondary sexual characteristics, such as voice change and facial hair.

**tetanus** (lockjaw) Life-threatening disease caused by the toxin secreted by the anaerobic bacterium *Clostridium tetani*. The symptoms are muscular spasms and rigidity of the jaw, which then spreads to other parts of the body, culminating in convulsions and death. The disease is treated with antitetanus toxin and ANTIBIOTICS.

◀ *Testosterone Polarized light micrograph of testosterone crystals. Testosterone may be used in hormone replacement therapy to treat delayed puberty in boys, hypogonadism, and some types of impotence. In women, it may be effective in the treatment of breast cancer.*

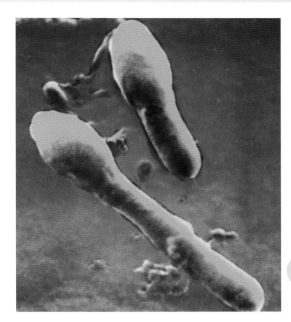

▶ *Tetanus False-color SEM of* Clostridium tetani, *the species of gram-positive, spore-forming, rodlike bacteria that causes tetanus. The spherical endospore appears as a swelling on top end of each bacterium. One resting endospore is produced from one vegetative bacterial cell and germinates to give rise to a single cell (i.e. the process is not one of multiplication). C. tetani exists harmlessly in the intestines of man and animals but becomes pathogenic (causes tetanus) when introduced through a wound.*
*Magnification: ×4,700.*

**T**

**tetracyclines** One of a group of broad-spectrum ANTIBIOTICS that are effective against a wide range of bacterial infections.

**thalassemia** (Cooley's anemia) Group of hereditary disorders characterized by abnormal bone marrow and erythrocytes (red blood cells). The predominant symptom is ANEMIA, requiring frequent blood transfusions.

**thalidomide** Drug originally developed as a mild hypnotic, but whose use by women in early pregnancy until the early 1960s came to be associated with serious birth deformities. It is still manufactured for occasional use in the treatment of LEPROSY and ACQUIRED IMMUNE DEFICIENCY SYNDROME (AIDS).

**thrombophlebitis** Inflammation of the walls of veins associated with THROMBOSIS. It can occur in a woman's legs during pregnancy.

**thrombosis** Formation of a blood clot in an artery or vein. Besides causing loss of circulation to the area supplied by the blocked vessel, thrombosis carries the risk of EMBOLISM.

**thrush** (candidiasis) Fungal infection of the mucous membranes, usually of the mouth but also sometimes of the vagina. Caused by the fungus *Candida albicans*, it is sometimes seen in people taking broad-spectrum ANTIBIOTICS.

**thyroxine** Hormone secreted by the thyroid gland. It contains iodine and helps regulate the rate of metabolism; it is essential for normal growth and development. *See also* pages 56–59

**tic** Sudden and rapidly repeated muscular contraction, limited to one part of the body, especially the face.

**tomography** Technique of X-RAY photography in which details of only a single slice or plane of body tissue are shown.

**tonsillitis** Acute or chronic inflammation of the tonsils caused by bacterial or viral

T

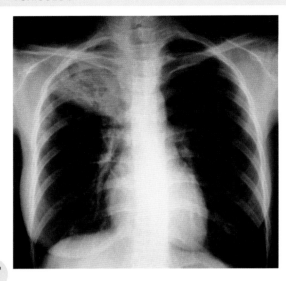

◄ *Tuberculosis* Colored chest X-ray of the lungs of a patient suffering from tuberculosis of the lung. The upper lobe of the lung (red, at left) is affected showing opacities (green). Healthy lung fields are black; the heart is at center. Tuberculosis is a disease in which the lungs develop dense opacities marking the infected sites. The infection is passed from person to person in airborne droplets. Symptoms include coughing, fever, chest pain, and shortness of breath. It is diagnosed by chest X-ray and sputum tests, and is treated with antibiotic drugs.

infection. It is signaled by fever, sore throat, and difficulty in swallowing. Chronic tonsillitis is often treated by surgical removal of the tonsils (tonsillectomy).

**toxicology** Study of poisonous substances and their effects on living things.

**toxic shock syndrome** Potentially fatal condition in which there is a dangerous drop in blood pressure and rapid onset of fever, diarrhea, vomiting, and muscular pains. It is caused by BLOOD POISONING (septicemia) arising from toxins put out by bacteria that normally reside in the body without causing harm. The syndrome is most often seen in young women using tampons during menstruation.

**toxin** Poisonous substance produced by a living organism. The unpleasant symptoms of many bacterial diseases are due to the release of toxins into the body by BACTERIA. Many molds, some larger fungi, and seeds of some higher plants produce toxins. The venom of many snakes contains powerful toxins. *See also* SNAKEBITE

**toxoplasmosis** Disease caused by the protozoan *Toxoplasma gondii*, which is transmitted from animals to human beings. It produces symptoms that are generally mild and flulike in adults, but it can damage the nervous system, eyes, and internal organs.

**trace elements** Chemical elements that are essential to life but normally obtainable from the DIET only in small quantities. They are essential to the reactions of enzymes and hormones.

**tracer, radioactive** Radioactive substance that is introduced into the body so that its progress can be tracked by special diagnostic equipment. This technique may help in the diagnosis of conditions such as thyroid disease.

**tracheotomy** (tracheostomy) Surgical procedure in which an incision is made through the skin into the trachea to allow insertion of a tube to facilitate breathing. It is done either to bypass any disease or damage in the trachea, or to safeguard the airway if a patient has to spend a long time on a mechanical ventilator. *See also* pages 74–75

**trachoma** Chronic eye infection caused by the microorganism *Chlamydia trachomatis*, characterized by inflammation of the cornea with the formation of pus. A disease of dry, tropical regions, it is the major cause of blindness in the developing world.

**tranquilizer** Drugs prescribed to reduce anxiety or tension and generally for their calming effect. They are used to control the symptoms of severe mental disturbance, such as schizophrenia or manic depression. They are also prescribed to relieve depression. Prolonged use of tranquilizers can produce a range of unwanted side effects.

**transfusion, blood** *See* BLOOD TRANSFUSION

**transplant** Surgical operation to introduce organ or tissue from one person (the donor) to another (the recipient); it may also refer to the transfer of tissues from one part of the body to another, as in grafting of skin or bone. Major transplants are performed to save the lives of patients facing death from end-stage organ disease. Organs routinely transplanted include the kidneys, heart, lungs, liver, and pancreas. Experimental work continues on some other procedures, including small-bowel grafting. Many other tissues are commonly grafted, including heart valves, bone, and bone marrow. The oldest transplant procedure is corneal grafting, which restores the sight of one or both eyes. In 1967, Christiaan Barnard performed the first successful heart transplant operation. Most transplant material is acquired from dead people, although kidneys, part of the liver, bone marrow, and corneas may be taken from living donors.

**trauma** Any injury or physical damage caused by some external event such as an accident or assault. In psychiatry, the term is applied to an emotional shock or harrowing experience.

▶ *Tumor A tumor is a swelling composed of cells independent of the body's control mechanism, so that they rapidly divide, invade, and kill surrounding tissue. Cross-sections of a healthy (A) and a diseased liver (B) are shown. The tumor shows a loss of cellular and structural differentiation and is unlike the tissue of origin.*

**tropical diseases** Any of a number of diseases predominantly associated with tropical climates. Major ones are MALARIA, LEISHMANIASIS, SLEEPING SICKNESS, FILARIASIS, and SCHISTOSOMIASIS (bilharzia). The infectious agents of tropical diseases include viruses, bacteria, protozoa, fungi, and worms of various kinds. Many of these disease microbes are spread by insect vectors such as mosquitoes.

**trypsin** Digestive enzyme secreted by the pancreas. It is secreted in an inactive form that is converted into active trypsin by an enzyme in the small intestine. It breaks down peptide bonds on the amino acids lysine and arginine. *See also* pages 88–89

**tuberculosis** (TB) Infectious disease caused by the bacillus *Mycobacterium tuberculosis*. It most often affects the lungs (pulmonary tuberculosis), but may involve the bones and joints, skin, lymph nodes, intestines, and kidneys. One-third of the world's population is infected, and up to 5% of those infected eventually develop TB. Poor urban living conditions have lead to a resurgence of the disease in the United States and much of Europe, where previously it had been in decline. The BCG vaccine against TB developed in the 1920s, and streptomycin, the first effective treatment drug, became available in 1944. However, the bacillus shows increasing resistance to drugs and some strains are multiresistant.

**tumor** Any uncontrolled, abnormal proliferation of cells, often leading to the formation of a lump. Tumors are classified

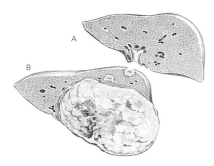

T

as benign (noncancerous) or malignant. *See also* CANCER

**Turner's syndrome** Hereditary condition in females, in which there is only one X-chromosome instead of two. It results in short stature, infertility, and developmental defects.

**typhoid fever** Acute, sometimes epidemic communicable disease of the digestive system. Caused by *Salmonella typhi*, which is transmitted in contaminated water or food, it is characterized by bleeding from the bowel and enlargement of the spleen. Symptoms include fever, headache, constipation, sore throat, cough, and skin rash.

**typhus** Any of a group of infectious diseases caused by rickettsiae (small bacteria) and spread by parasites of the human body such as lice, fleas, ticks, and mites. **Epidemic** typhus, the result of infection by *Rickettsia prowazekii*, is the most serious manifestation. Associated with dirty, overcrowded conditions, it is mainly seen during times of war or famine.

**T**

**U**

**V**

**ulcer** Any persistent sore or lesion on the skin or on a mucous membrane, often associated with inflammation. Ulcers may be caused by infection, chemical irritation or mechanical pressure.

**urethritis** Inflammation of the urethra. It is usually due to a SEXUALLY TRANSMITTED DISEASE (STD) but may also arise from infection.

**urine** Fluid filtered out from the bloodstream by the KIDNEY. It consists mainly of water, salts, and waste products such as urea. From the kidneys it passes through the ureters to the bladder for voiding by way of the urethra. *See also* pages 96–97

**urology** Medical specialty concerned with the diagnosis and treatment of diseases of the urinary tract in women and of the urinary and reproductive systems in men.

**urticaria** *See* HIVES

**vaccination** Injection of a VACCINE in order to produce IMMUNITY against a disease. In many countries, children are routinely vaccinated against infectious diseases.

**vaccine** Agent used to give IMMUNITY against various diseases without producing symptoms. A vaccine consists of modified disease organisms, such as live, weakened virus, or dead ones that are still able to induce the production of specific antibodies within the blood. *See also* ANTI-BODY; IMMUNE SYSTEM; VIRUS

**Valium** Proprietary name for diazepam, a sedative drug in the BENZODIAZEPINE group. It is used in the treatment of anxiety, muscle spasms, and epilepsy.

**varicose vein** Condition where a vein becomes swollen and distorted. Varicose veins can occur anywhere in the body, but are commonly found in the legs.

**vasectomy** Operation to induce male sterility, in which the tube (*vas deferens*) carrying sperm from the testes to the penis is cut. A vasectomy is a form of permanent CONTRACEPTION, although in some cases the operation is reversible.

**vasoconstrictor** Any substance that causes constriction of blood vessels and, therefore, decreased blood flow. Examples include NOREPINEPHRINE, angiotensin, and the hormone vasopressin (also known as antidiuretic hormone).

**vasodilator** Any substance that causes widening of the blood vessels, permitting freer flow of blood. Vasodilator drugs are mostly used to treat HYPERTENSION and ANGINA.

**venereal disease (VD)** Any of a number of diseases transmitted through sexual contact, chief of which are SYPHILIS, GONORRHEA, and chancroid. **Syphilis** is caused by the bacterium *Treponema pallidum*. PENICILLIN and its derivatives can still cure syphilis in its early stages.

Gonorrhea is caused by the gonococcus bacterium and, if diagnosed early, may be treated with SULFONAMIDE DRUGS. *See also* SEXUALLY TRANSMITTED DISEASE (STD)

**verruca** In medicine, form of WART on the sole of the foot, which is painful because it is forced to grow inward. Like other warts, it is due to infection with the human papillomavirus.

**vertigo** Dizziness, often accompanied by nausea. Due to disruption of the sense of balance, it may be produced by ear disorder, reduced flow of blood to the brain caused by altitude, emotional upset, or spinning rapidly.

**virology** Study of VIRUSES. The existence of viruses was established (1892) by D. Ivanovski, a Russian botanist, who found that the causative agent of tobacco mosaic disease could pass through a porcelain filter impermeable to BACTERIA. The introduction of the electron micro-

▶ *Virus Bacteriophages are viruses that infect bacteria. Bacteriophage lambda (A) is a virus that infects E. coli bacteria in the intestine. The DNA (1) is stored in a polyhedral protein head (2) attached to a hollow tail (3) with a single tail fiber (4). The phage attaches itself to a bacterium (B) at the tail (1), and injects its DNA (2). This can result in an infection that causes the cell to lyse (break open), releasing replica phages. During such an infection, the phage DNA remains separate from the bacterial DNA (3), and uses the cell's enzymes to synthesize the proteins that make up the new phages (4). The phage DNA replicates (5). In the process, the bacterial DNA is used and, by the time the cell lyses (6), it may be destroyed. The injection of phage DNA may result in a lysogenic infection (7), when the phage DNA becomes part of the bacterial chrmosome (8). Cell division (9) then produces numerous replicas of the phage DNA. During the course of the lysogenic growth, damage to the cell may result in a lytic infection (10) by causing the phage DNA to be ejected from the bacterial chromosome.*

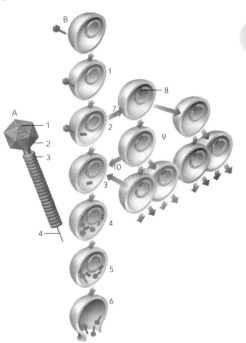

**V**

scope in the 1940s made it possible to view viruses. *See also* VIRUS

**virus** Submicroscopic infectious organism. Viruses vary in size from ten to 300 nanometers, and contain only genetic material in the form of DNA or RNA. Viruses are incapable of independent existence: they can grow and reproduce only when they enter another cell, such as a bacterium or animal cell, because they lack energy-producing and protein-synthesizing functions. When they enter a cell, viruses subvert the host's metabolism so that viral reproduction is favored. Control of viruses is difficult because harsh measures are required to kill them. The animal body has, however, evolved some protective measures, such as production of INTERFERON and of antibodies directed against specific viruses. Where the specific agent can be isolated, VACCINES can be developed, but some viruses change so rapidly that vaccines become ineffective. *See artwork*, page 184. *See also* ANTIBODY; IMMUNE SYSTEM

**vomiting** Act of bringing up the contents of the stomach by way of the mouth. Vomiting is a reflex mechanism that may be activated by any of a number of stimuli, including dizziness, pain, gastric irritation, or shock. It may also be a symptom of serious disease.

 **W**

**wart** Raised and well-defined small growth on the outermost surface of the skin, caused by the human papillomavirus. It is usually painless unless in a pressure area, as with a VERRUCA.

**whooping cough** (pertussis) Acute, highly contagious childhood respiratory disease. It is caused by the bacterium *Bordetella pertussis* and is marked by spasms of coughing, followed by a long drawn intake of air, or 'whoop.' It is frequently associated with vomiting and severe nose bleeds. Immunization reduces the number and severity of attacks.

**World Health Organization (WHO)** Intergovernmental organization, a specialized agency of the United Nations (UN). Founded in 1948, it collects and

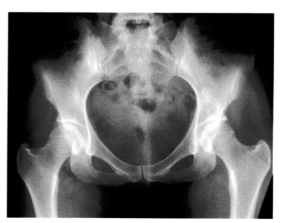

◀ **X-ray** *False-color anteroposterior X-ray of a normal adult, female pelvis. The pelvis is formed by the sacrum, coccyx, and two innominate (hip) bones. Here, the sacrum is at top center, articulating with each innominate bone at the sacro iliac joints. Left and right hip bones meet at the symphysis pubis (bottom). Also visible is the articulation of the rounded head of each femur (thigh bone) with each hip socket.*

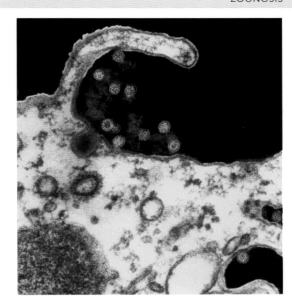

▶ *Yellow fever* Colored TEM of yellow fever viruses (yellow/red, at center) on the surface of an infected cell. The cell is white/brown. This virus occurs in tropical Africa, and Central and South America. It is transmitted to humans when an infected female mosquito Aedes aegypti bites a human. Yellow fever is so-called because it may cause yellowing of the skin due to jaundice. About 10% of victims die.
Magnification: x50,000.

shares medical and scientific information and promotes the establishment of international standards for drugs and vaccines. WHO has made major contributions to the prevention of diseases such as malaria, polio, leprosy, and tuberculosis, and the eradication of smallpox. Its headquarters are in Geneva, Switzerland.

**X-ray** Electromagnetic radiation of shorter wavelength (or higher frequency) than visible light, produced when a beam of electrons hits a solid target. Wilhelm Röntgen discovered X-rays in 1895. They are normally produced for scientific use in X-ray tubes. Because they are able to penetrate matter that is opaque to light, X-rays are used to investigate inaccessible areas, especially of the body.

**yaws** (frambesia) Contagious skin disease found in the humid tropics. It is caused by a spirochete (*Treponema pertenue*), related to the organism causing SYPHILIS. Yaws is not a SEXUALLY TRANSMITTED DISEASE (STD), but rather is transmitted by flies and by direct skin contact with the sores. It may go on to cause disfiguring bone lesions.

**yellow fever** Acute, infectious disease marked by sudden onset of headaches, fever, muscle and joint pain, jaundice, and vomiting; the kidneys and heart may also be affected. It is caused by a VIRUS transmitted by mosquitoes in tropical and subtropical regions. It may be prevented by vaccination.

**zoonosis** Any infection or infestation of vertebrates that is transmissible to human beings.

# Index

Page numbers in *italic*
type refer to illustrations
or their captions; **bold**
numbers refer to major
references.